HARCOURT Math

Benchmark Assessments for Grade 4 California Mathematics Standards

TEACHER'S EDITION

Harcourt

Orlando   Austin   Chicago   New York   Toronto   London   San Diego

Visit *The Learning Site!*
www.harcourtschool.com

Copyright © by Harcourt, Inc.

All rights reserved. No part of this publication may be reproduced or transmitted in any form or by any means, electronic or mechanical, including photocopy, recording, or any information storage and retrieval system, without permission in writing from the publisher.

Teachers using HARCOURT MATH may photocopy Copying Masters in complete pages in sufficient quantities for classroom use only and not for resale.

HARCOURT and the Harcourt Logo are trademarks of Harcourt, Inc.

Printed in the United States of America

ISBN 0-15-335856-4

1 2 3 4 5 6 7 8 9 10   073   10 09 08 07 06 05 04 03 02

CONTENTS

- Overview iv
- Using the Think Along Pages v
- Using Basic Skills Tests v
- Using Benchmark Practice Pages v
- Using Benchmark Assessments vi
- Using Practice Tests vi
- Using Assessment Summary and Prescriptions Charts vi
- Think Alongs T-1–T-22
- Basic Skills Practice Tests 1–10
- Benchmark Practice 1 11–22
- Benchmark Assessment 1 23–30
- Benchmark Practice 2 31–42
- Benchmark Assessment 2 43–50
- Benchmark Practice 3 51–62
- Benchmark Assessment 3 63–70
- Practice Test 1 71–78
- Practice Test 2 79–86
- Assessment Summary and Prescriptions Charts 87–98
- Correlations 99–103

Overview

Harcourt Math Benchmark Assessments for California Mathematics Standards has been designed to help students review and reinforce their knowledge of the California Mathematics Standards.

- Think Along pages are designed for use at school or at home to stimulate development of problem-solving and test-taking skills.

- Basic Skills Practice Tests help students develop fluency with basic facts and other mental math skills.

- Benchmark Practice Pages that focus on the California Mathematics Standards covered within a three-unit content cluster are presented in mixed formats prior to the benchmark assessment for the same content cluster.

- Benchmark Assessments provide cumulative review at the end of each three-unit content cluster in *Harcourt Math*. Items are presented in mixed formats, and each item is labeled with the standard being reviewed.

- Two Practice Tests provide an opportunity to pre-test and post-test the math content for an entire year of study.

- Benchmark Assessment Summary and Prescriptions chart is provided for each Benchmark Assessment. The chart includes the California Mathematics Standards, item numbers assessing each standard, *Harcourt Math* lesson numbers where the standard is covered, and prescriptions for intervention as needed.

- Answer Keys are provided in the Teacher's Edition at point of use.

Using the Think Along Pages

Each Think Along page presents two or more problems for students to solve. Each set of problems is aligned to a key standard at the grade level. Prompts are provided to facilitate discussion of solution strategies and review key vocabulary.

You may want to use one page each week in the classroom OR send pages home. Families will then have the opportunity to become familiar with the content expectations at the grade level and can help students achieve success.

Using Basic Skills Practice Tests

The Basic Skills Practice Tests are designed to provide quick practice with key computational skills at each grade. You may want to make overhead transparencies of these pages. Have students complete two rows of problems per day over a period of weeks.

Short daily review can be an effective method for developing memorization skills that lead to fluency with basic skills computation.

Using Benchmark Practice Pages

The Benchmark Practice pages review the California Mathematics Standards that are taught in each three-unit content cluster in *Harcourt Math*. The pages are organized by strand and provide practice in mixed formats.

These pages provide students with focused practice on number sense; algebra and functions; measurement and geometry; statistics, data analysis, and probability; and mathematical reasoning. These same standards are then provided in a follow-up Benchmark Assessment for the three-unit content cluster.

Using the Benchmark Assessments

Each Benchmark Assessment is a cumulative review of the California Mathematics Standards covered in each three-unit cluster in *Harcourt Math*. These items are presented in mixed formats including multiple-choice items, free-response items, and write-about-it items where students provide explanations for their answer.

Use these assessments after you complete every three units to help students retain previously learned skills. Each item references the standard covered. After students have completed the assessment, refer to the Benchmark Assessment Summary and Prescriptions chart in the back of this book to provide intervention and remediation as needed.

Using the Practice Tests

Two Practice Tests cover the entire year's curriculum. You may want to use these as pre-test and post-test.

Use Practice Test 1 as a pre-test to set a baseline. Use Practice Test 2 as a post-test to document each student's progress.

Using the Benchmark Assessment Summary and Prescriptions Charts

The Benchmark Assessment Summary and Prescriptions chart provides you with a blueprint of each assessment. Included in the summary chart are the California Mathematics Standards being assessed, the item numbers in the assessment covering each standard, the lesson numbers in *Harcourt Math* where that standard is taught, and suggested intervention and review options.

These review options offer a variety of materials and strategies for helping students achieve success. Prescriptions include references to *Harcourt Math* Reteach and Practice Workbook pages, Extra Practice sets in the Pupil's Edition, Alternative Teaching Strategies in the Teacher's Edition, and Intervention Strategy Skills on the Intervention CD-ROM or the Intervention Skill Cards.

NS 1.1
Read and write whole numbers in the millions.

Read whole numbers in the millions.

1 What is the standard form for fifty-eight million, two hundred nine thousand, eighty-six?

A 58,290,860
B 58,209,860
C 58,209,086
D 58,029,086

SOLUTION/EXPLANATION

To solve this problem, write the digits for each period. Then write the periods in order.

Millions	Thousands	Ones
58,	209,	086

fifty-eight million → 58 millions

two hundred nine thousand → 209 thousands

eighty-six → 86 ones

The correct choice is: Ⓐ Ⓑ ● Ⓓ

Write whole numbers in the millions.

2 Which number is 100,000 greater than 6,529,200?

A six million, five hundred twenty-nine thousand, two hundred ten

B six million, five hundred twenty-nine thousand, three hundred

C six million, five hundred thirty-nine thousand, two hundred

D six million, six hundred twenty-nine thousand, two hundred

SOLUTION/EXPLANATION

To solve this problem add the two numbers.

```
  6,529,200
+   100,000
  ─────────
  6,629,200
```

Then write the number in word form from left to right by period.

six million,

six hundred twenty-nine thousand,

two hundred

The correct choice is: Ⓐ Ⓑ Ⓒ ●

Benchmark Assessment

T-1

NS 1.2

Order and compare whole numbers and decimals to two decimal places.

Order and compare whole numbers.

1 Which number sentence is **not** true?

A 18,402 < 20,999
B 51,042 > 34,012
C 21,007 = 21,007
D 50,340 < 45,789

SOLUTION/EXPLANATION

To solve this problem, think about the position of each number on a number line. Lesser numbers are to the left of greater numbers. Next to each sentence write *true* or *not true*.

18,402 < 20,999 **True**

51,042 > 34,012 **True**

21,007 = 21,007 **True**

50,340 < 45,789 **Not True**

On a number line 45,789 is to the left of 50,340, so it is less than 50,340. The number sentence 50,340 < 45,789 is not true.

The correct choice is: Ⓐ Ⓑ Ⓒ **Ⓓ**

Order and compare decimals to two decimal places.

2 Which is the **least** number?

A 0.03
B 0.33
C 0.30
D 3.03

SOLUTION/EXPLANATION

One way to compare these decimal numbers is to write each as a fraction with a denominator of 100. Then look for the least numerator.

$0.03 = \frac{3}{100}$

$0.33 = \frac{33}{100}$

$0.30 = \frac{30}{100}$

$3.03 = \frac{303}{100}$

The least numerator is 3, so 0.03 is the least decimal among these choices.

The correct choice is: **Ⓐ** Ⓑ Ⓒ Ⓓ

NS 1.3

Round whole numbers through the millions to the nearest ten, hundred, thousand, ten thousand, or hundred thousand.

THINK ALONG

Round whole numbers to a given unit.

1 Sandra collected 2,344 sports cards. How many cards is this rounded to the nearest hundred?

    A 3,000
    B 2,400
    C 2,350
    D 2,300

SOLUTION/EXPLANATION

To round 2,344 to the nearest hundred, underline the digit in the hundreds place. Then look at the digit to the right of the underlined digit.

    2,344 → 2,300

Since it is 5 or less, the underlined digit stays the same. Rewrite the digits to the right of the underlined digit with zeros.

The correct choice is: Ⓐ Ⓑ Ⓒ **Ⓓ**

2 A recent report stated that Broward County, FL, has a population of 1,382,983 people. Round this number to the nearest hundred thousand.

    A 1,400,000
    B 1,383,000
    C 1,380,000
    D 1,300,000

SOLUTION/EXPLANATION

To round 1,382,983 to the nearest hundred thousand, underline the digit in the hundred thousands place. Then look at the digit to the right of the underlined digit.

    1,382,983 → 1,400,000

Since it is 5 or greater, increase the underlined digit by one. Rewrite the digits to the right of the underlined digit with zeros.

The correct choice is: **Ⓐ** Ⓑ Ⓒ Ⓓ

Benchmark Assessment

NS 1.4

Decide when a rounded solution is called for and explain why such a solution may be appropriate.

Decide when a rounded solution is called for.

1 The attendance at a soccer match was 51,534 people. The newspaper reported the number to the nearest thousand. What was the reported attendance?

A about 50,000
B about 51,000
C about 52,000
D about 60,000

SOLUTION/EXPLANATION

To round 51,534 to the nearest thousand, first underline the thousands digit. Then check the digit to its right. Since this digit is 5 or greater, you need to increase the underlined digit by 1. The nearest thousand is 52,000.

The correct choice is: Ⓐ Ⓑ

2 For which of the following is only a rounded amount needed?

A the amount of money needed to pay for a sandwich
B the number of cookies needed to feed everyone two cookies each
C the number of berries in the average pint of berries
D the amount of sugar required to bake a cake

SOLUTION/EXPLANATION

To find the appropriate answer, find the response that best represents a rounded amount. In response A, a rounded amount may not be appropriate. In responses B and D, a rounded amount may allow for too few or too many. The number of berries in the average pint of berries will change depending on the size of the berries, so a rounded amount is appropriate.

The correct choice is: Ⓐ Ⓑ

Benchmark Assessment

NS 1.8

Use concepts of negative numbers (e.g., on a number line, in counting, in temperature, in "owing").

Use concepts of negative numbers.

1 Which number is least?

A 8
B ⁻6
C 4
D ⁻2

SOLUTION/EXPLANATION

To solve this problem, locate the point for each number on a number line. The least number is the one farthest to the left.

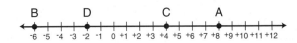

The correct choice is: Ⓐ **Ⓑ** Ⓒ Ⓓ

2 Which number sentence is **not** true?

A ⁺6 > ⁻6
B ⁻2 > ⁻4
C ⁺7 < ⁺12
D ⁻4 > ⁻3

SOLUTION/EXPLANATION

One way to solve this problem is to draw a number line.

Look at each number sentence. Then compare the location of each pair of numbers. Recall that numbers along the number line increase from left to right. So, ⁻4 > ⁻3 is not true, because ⁻4 is to the left of ⁻3.

The correct choice is: Ⓐ Ⓑ Ⓒ **Ⓓ**

Benchmark Assessment

 **NS 1.9**

THINK ALONG

Identify on a number line the relative position of positive fractions, positive mixed numbers, and positive decimals to two decimal places.

Identify on a number line the relative position of fractions.

1 Which is the equivalent fraction for point A?

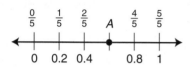

A $\frac{1}{5}$

B $\frac{2}{5}$

C $\frac{3}{5}$

D $\frac{4}{5}$

SOLUTION/EXPLANATION

Look at the number line. It is marked into fifths. Since A is three marks to the right of zero, it is at $\frac{3}{5}$.

The correct choice is: Ⓐ Ⓑ ●Ⓒ Ⓓ

Identify on a number line the relative position of mixed numbers.

2 Which point represents $3\frac{5}{6}$?

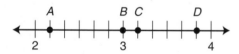

A A

B B

C C

D D

SOLUTION/EXPLANATION

Look at the number line. It is marked into sixths. To locate $3\frac{5}{6}$, count five sixths to the right of 3. This is point D.

The correct choice is: Ⓐ Ⓑ Ⓒ ●Ⓓ

© Harcourt

T-6

Benchmark Assessment

NS 3.0

Students solve problems involving addition, subtraction, multiplication, and division of whole numbers and understand the relationships among the operations.

Solve problems involving addition and subtraction.

1 The table shows the number of caps collected in one week by 3 classes at Lincoln School for a project. How many caps were collected by the three classes?

BOTTLE CAP COLLECTIONS

Class	Number Collected
Ms. Chin	59
Mr. Smith	141
Ms. Howard	327

A 1,527 C 1,058
B 1,258 D 527

SOLUTION/EXPLANATION

To solve the problem, add the numbers of caps. Be sure to align the ones place in all the numbers.

```
   59
  141
+ 327
-----
  527
```

The correct choice is: Ⓐ Ⓑ Ⓒ **Ⓓ**

2 Taleah has collected exactly 1,000 baseball cards in all. Last year she had only 449 cards. How many cards did she collect this year?

A 1,449
B 651
C 561
D 551

SOLUTION/EXPLANATION

You need to find the difference in the numbers of cards. Write a subtraction problem. Use addition to check.

```
 1,000      Check:     551
-  449                + 449
-----                 -----
   551                1,000
```

The correct choice is: Ⓐ Ⓑ Ⓒ **Ⓓ**

Solve problems involving multiplication.

3 Sam is stocking the egg case at the store. He has 16 cartons of eggs with 18 eggs in each carton. How many eggs is that?

A 648
B 288
C 128
D 32

SOLUTION/EXPLANATION

Estimate the product. Since 20 × 20 = 400, choices C and D are unreasonable. To check that choice B is correct, multiply.

```
    18
 ×  16
-----
   108    (6 × 18)
   180    (10 × 18)
-----
   288
```

The correct choice is: Ⓐ **Ⓑ** Ⓒ Ⓓ

Benchmark Assessment

T-7

NS 3.1

Demonstrate an understanding of, and the ability to use, standard algorithms for the addition and subtraction of multidigit numbers.

Use standard algorithms for the addition of multidigit numbers.

1 Sarah spent $2,395 on new kitchen appliances and $9,318 on new cabinets. How much did she spend in all?

A $6,923
B $9,713
C $11,703
D $11,713

SOLUTION/EXPLANATION

To solve the problem, add the two numbers together. Be sure to align the digits in the correct place value.

```
   2,395
 + 9,318
 ------
  11,713
```

The correct choice is: Ⓐ Ⓑ Ⓒ **Ⓓ**

Use standard algorithms for the subtraction of multidigit numbers.

2 The *Indianapolis Star* newspaper sold 231,423 papers one Sunday. The *Kansas City Star* newspaper sold 290,650 papers. How many more papers did the Kansas City newspaper sell than the Indianapolis newspaper?

A 57,227   C 60,000
B 59,227   D 522,073

SOLUTION/EXPLANATION

To solve the problem, subtract. Be sure to align the digits in the correct place value.

```
  290,650
- 231,423
 -------
   59,227
```

The correct choice is: Ⓐ **Ⓑ** Ⓒ Ⓓ

3 The Horse Club had $8,900. It spent $3,457 to paint the barn and $2,400 to run the annual horse show. How much money is left?

A $1,057
B $3,043
C $5,857
D $14,757

SOLUTION/EXPLANATION

There are several ways to solve this problem. One method is to add the cost of painting the barn and the cost of the horse show.

```
  $3,457
+  2,400
 ------
  $5,857
```

Next subtract the costs from the amount of money the club had.

```
  $8,900
-  5,857
 ------
  $3,043
```

The correct choice is: Ⓐ **Ⓑ** Ⓒ Ⓓ

T-8

Benchmark Assessment

> THINK ALONG

 NS 3.2

Demonstrate an understanding of, and the ability to use, standard algorithms for multiplying a multidigit number by a two-digit number and for dividing a multidigit number by a one-digit number; use relationships between them to simplify computations and to check results.

Use the standard algorithm for multiplying a multidigit number by a two-digit number.

1 What is the product?

28 × $3,049

- **A** $9,147
- **B** $31,290
- **C** $85,372
- **D** $85,381

SOLUTION/EXPLANATION

First estimate the product by rounding.
30 × 3,000 = 90,000

Choices A and B are unreasonable. To solve the problem, you can multiply.

```
    3,049
  ×    28
   24,392   (8 × 3,049)
   60,980   (20 × 3,049)
   85,372
```

The correct choice is: Ⓐ Ⓑ ● Ⓓ

Use the standard algorithm for dividing a multidigit number by a one-digit number.

2 4)2,025

- **A** 42
- **B** 55
- **C** 56 r1
- **D** 506 r1

SOLUTION/EXPLANATION

First estimate using compatible numbers.

2,025 ÷ 4
  ↓   ↓
2,000 ÷ 4 = 500

Choices A, B, and C are unreasonable. Divide to find the correct answer.

```
     506 r1
4)2,025
  −20
   02
  − 0
    25
   −24
     1
```

The correct choice is: Ⓐ Ⓑ Ⓒ ●

Benchmark Assessment

NS 3.3
Solve problems involving multiplication of multidigit numbers by two-digit numbers.

Solve problems involving multiplication of multidigit numbers by two-digit numbers.

---

1 Ellen's heart rate is 65 beats per minute. In 60 minutes her heart beats about 3,900 times. If Ellen's heart continues beating at the same rate, how many times would her heart beat in 24 hours?

   A 1,440 times
   B 23,400 times
   C 93,600 times
   D 5,616,000 times

SOLUTION/EXPLANATION

One strategy is to eliminate choices that are unreasonable. Estimate the product by rounding 24 (hours) to the nearest 10, and 3,900 to the nearest thousand.

24 × 3,900

20 × 4,000 = 80,000

So choices A, B, and D are unreasonable.

The correct choice is: Ⓐ Ⓑ ● Ⓓ

---

2 Mr. Brandon is offered a job that pays $3,560 each month. How much money will he make in a year?

   A $42,720
   B $41,720
   C $20,680
   D $10,680

SOLUTION/EXPLANATION

One strategy is to eliminate choices that are unreasonable. Estimate the product by rounding $3,560 to the nearest thousand dollars.

12 × 4,000 = 48,000

So choices C and D are unreasonable. To solve the problem, multiply.

```
      3,560
   ×     12
      7,120   (2 × 3,560)
     35,600   (10 × 3,560)
     42,720
```

The correct choice is: ● Ⓑ Ⓒ Ⓓ

NS 3.4
Solve problems involving division of multidigit numbers by one-digit numbers.

Solve problems involving division of multidigit numbers by one-digit numbers.

1 The car wash fundraiser made $2,296. The money is split evenly between 8 clubs. How much will each club receive?

  A $28.70
  B $247.00
  C $267.00
  D $287.00

SOLUTION/EXPLANATION

To solve this problem, think about a compatible number for $2,296.

2,296 ÷ 8
   ↓   ↓
2,400 ÷ 8 = 300

Answer choice A is unreasonable. Divide to find which choice is correct.

```
      287
   8)2,296
    -16
     ‾‾‾
      69
     -64
     ‾‾‾
       56
      -56
      ‾‾‾
        0
```

Each club receives $287.

The correct choice is: Ⓐ Ⓑ Ⓒ **Ⓓ**

2 There are 285 students going to the museum. Each car can hold 6 students. How many cars are needed?

  A 1,710 cars
  B 48 cars
  C 47 cars
  D 30 cars

SOLUTION/EXPLANATION

To solve this problem, find a compatible number for 285.

285 ÷ 6
  ↓   ↓
300 ÷ 6 = 50

Answer choices A and D are unreasonable. Divide to find which choice is correct.

Forty-seven cars can carry 282 students. One more car can carry the remaining 3 students. So, 48 cars are needed to carry 285 students.

The correct choice is: Ⓐ **Ⓑ** Ⓒ Ⓓ

Benchmark Assessment

NS 4.2

Know that numbers such as 2, 3, 5, 7, and 11 do not have any factors except 1 and themselves and that such numbers are called prime numbers.

Understand the meaning of prime numbers.

1 Which set of numbers contains only prime numbers?

  A 3, 7, 15
  B 5, 17, 39
  C 7, 11, 19
  D Not Given

SOLUTION/EXPLANATION

To solve this problem, find the factors of each number. If all the numbers have only 1 and the number itself as factors, then the set of numbers is prime.

  A 3 (1, 3; prime), 7 (1, 7; prime), 15 (1, 3, 5, 15; composite)
  B 5 (1, 5; prime), 17 (1, 17; prime), 39 (1, 3, 13, 39; composite)
  C 7 (1, 7; prime), 11 (1, 11; prime), 19 (1, 19; prime)

Response C is the only set of numbers that contains only prime numbers.

The correct choice is: Ⓐ Ⓑ ● Ⓓ

2 Which of the following numbers is **not** a prime number?

  A 25
  B 29
  C 37
  D 43

SOLUTION/EXPLANATION

Numbers that have only 1 and the number itself as factors are **prime numbers.** If you are not sure whether a number is prime, divide by divisors that are less than the number. Recall your fact families or use a calculator. For example, none of the numbers from 2 to 36 divides 37 evenly.

The numbers 43, 37, and 29 are prime. Since 25 is divisible by 1, 5, and 25, it is a not a prime number.

The correct choice is: ● Ⓑ Ⓒ Ⓓ

Benchmark Assessment

AF 1.2
Interpret and evaluate mathematical expressions that now use parentheses.

Interpret and evaluate mathematical expressions that use parentheses.

THINK ALONG

---

1 What is the value of the expression?

200 + (178 − 21)

A 1
B 43
C 357
D 399

SOLUTION/EXPLANATION

First perform the operation in the parentheses.

200 + **(178 − 21)**
↓
200 + **157**
↓
357

The correct choice is: Ⓐ Ⓑ ● Ⓓ

---

2 What is the value of the expression?

(4 + 7) × (30 − 25)

A 15
B 55
C 189
D Not Given

SOLUTION/EXPLANATION

Do the operations within the parentheses first.

(4 + 7) × (30 − 25)
↓   ↓
11   ×   5
↓
55

The correct choice is: Ⓐ ● Ⓒ Ⓓ

Benchmark Assessment

T-13

AF 1.3

Use parentheses to indicate which operation to perform first when writing expressions containing more than two terms and different operations.

Use parentheses to indicate which operation to perform first when writing expressions containing more than two terms and different operations.

1 Which expression has a value of 27?

A $(81 - 9) \times 6$
B $81 - (9 \times 6)$
C $81 - (9 \div 6)$
D $(81 \div 9) - 6$

SOLUTION/EXPLANATION

To solve this problem, evaluate each expression. Begin with the operation in the parentheses.

$(81 - 9) \times 6$ → $72 \times 6 = 432$
$81 - (9 \times 6)$ → $81 - 54 = 27$

The correct choice is: Ⓐ **Ⓑ** Ⓒ Ⓓ

2 Ms. Clarke used a $20 bill to buy $8 worth of fruit and $9 worth of vegetables. Which expression shows how much money she has left?

A $(20 - 8) + 9$
B $20 - 8 + 9$
C $20 - (8 + 9)$
D $20 + 8 - 9$

SOLUTION/EXPLANATION

The problem says $8 and $9 were spent, so you can write $(8 + 9)$. This amount is subtracted from $20 so you can write
$20 - (8 + 9)$.

The correct choice is: Ⓐ Ⓑ **Ⓒ** Ⓓ

THINK ALONG

T-14   **Benchmark Assessment**

> **THINK ALONG**

AF 1.5

Understand that an equation such as $y = 3x + 5$ is a prescription for determining a second number when a first number is given.

Understand that an equation such as $y = 3x + 5$ is a prescription for determining a second number when a first number is given.

1 Which equation shows the rule for the table?

Input	z	2	3	4	5
Output	y	6	7	8	9

- **A** $z + 3 = y$
- **B** $z \times 3 = y$
- **C** $z + 4 = y$
- **D** $y \div 3 = z$

SOLUTION/EXPLANATION

Look at the table to see if you can find a pattern. Use a rule to describe the pattern. Test your rule using the numbers in the table.

The rule is "add 4." The equation $z + 4 = y$ shows this rule.

The correct choice is: Ⓐ Ⓑ Ⓒ Ⓓ

2 Which equation shows the rule for the table?

Input, z	Output, y
8	4
10	5
12	6
14	7

- **A** $z \times 3 = y$
- **B** $z \div 2 = y$
- **C** $z + y = 14$
- **D** $z - y = 2$

SOLUTION/EXPLANATION

Look at the table to see if you can find a pattern. Use a rule to describe the pattern. Test your rule using the numbers in the table.

The rule is "divide by 2." The equation $z \div 2 = y$ shows this rule.

The correct choice is: Ⓐ Ⓑ Ⓒ Ⓓ

Benchmark Assessment

AF 2.0

Students know how to manipulate equations.

Students know how to manipulate equations.

1 What number makes this equation true?

$3 + 8 + 5 = 11 + \square$

A 16
B 11
C 8
D 5

SOLUTION/EXPLANATION

You know that $3 + 8 = 11$ and that adding 5 to both sides of the equation will make another true equation. So you must add 5 to the right hand side of the equation.

Check: $3 + 8 + 5 = 11 + 5$.

The correct choice is: (A) (B) (C) **(D)**

2 Multiply both sides of the equation by 3. What are the new values?

$8 = 2 \times 4$

A $8 = 24$
B $11 = 3$
C $24 = 8$
D $24 = 24$

SOLUTION/EXPLANATION

Look at both sides of the equation. Multiply each side by 3.

$8 \rightarrow 8 \times 3 = 24$

$2 \times 4 \rightarrow 2 \times 4 \times 3 = 24$

The new values are $24 = 24$.

The correct choice is: (A) (B) (C) **(D)**

AF 2.1

Know and understand that equals added to equals are equal.

Know and understand that equals added to equals are equal.

1 Which coins **do not** make the value of the two sides equal?

   1 dime + 2 quarters = ___?___

   A 5 dimes, 10 pennies
   B 1 quarter, 2 dimes, 3 nickels
   C 10 nickels, 1 dime
   D 20 pennies, 1 quarter, 1 nickel

SOLUTION/EXPLANATION

The left side of the equation equals 60 cents. The value of the coins on the right side of the equation has to equal 60 cents as well. The value of Response D is 50 cents, so these coins do not balance the equation.

The correct choice is: Ⓐ Ⓑ Ⓒ **Ⓓ**

2 Complete to make the equation true.

   15 + 7 + 6 = ___?___ + 6

   A 15
   B 22
   C 24
   D 26

SOLUTION/EXPLANATION

Six has been added to both sides of the equation. You need to find a value equal to 15 + 7.

The correct choice is: Ⓐ **Ⓑ** Ⓒ Ⓓ

Benchmark Assessment   T-17

AF 2.2

Know and understand that equals multiplied by equals are equal.

Know and understand that equals multiplied by equals are equal.

1 Both sides of the first equation have been multiplied by 2 to form the second equation. Which coins complete the second equation?

8 pennies = 1 nickel + 3 pennies

8 pennies × 2 = ▨

- **A** 1 nickel + 3 pennies
- **B** 1 nickel + 6 pennies
- **C** 2 nickels + 3 pennies
- **D** 2 nickels + 6 pennies

SOLUTION/EXPLANATION

When you multiply the first equation by 2, you need to multiply all the pennies and nickels by 2.

1 nickel + 3 pennies
$\underline{\qquad\qquad\times 2}$
2 nickels + 6 pennies

The correct choice is: Ⓐ Ⓑ Ⓒ **Ⓓ**

2 Multiply both sides of this equation by 4. What are the new values?

6 + 9 = 3 × 5

- **A** 15 = 8
- **B** 60 = 60
- **C** 33 = 60
- **D** 42 = 15

SOLUTION/EXPLANATION

Multiply each side of the equation by 4.

6 + 9 → (6 + 9) × 4 = 60
3 × 5 → (3 × 5) × 4 = 60

The new values are 60 = 60.

The correct choice is: Ⓐ Ⓑ **Ⓒ** Ⓓ

MG 2.0

Students use two-dimensional coordinate grids to represent points and graph lines and simple figures.

Students use two-dimensional coordinate grids to represent points.

1 Which point is at (3,2)?

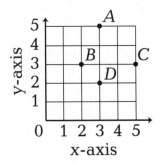

- **A** A
- **B** B
- **C** C
- **D** D

SOLUTION/EXPLANATION

The coordinates tell you how far to move from (0,0). To get to (3,2), count 3 units to the right and 2 units up from (0,0). The point where you stop is labeled D.

The correct choice is: Ⓐ Ⓑ Ⓒ ●

2 Which ordered pair names point Q?

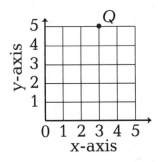

- **A** (2,3)
- **B** (3,2)
- **C** (5,3)
- **D** (3,5)

SOLUTION/EXPLANATION

Point Q is located at (3,5). To locate the coordinates of a point, start at 0. On the horizontal axis, count the number of units to the point. Point Q is located 3 units to the right (the first coordinate), and 5 units up (the second coordinate).

The correct choice is: Ⓐ Ⓑ Ⓒ ●

Benchmark Assessment

MG 2.1

Draw the points corresponding to linear relationships on graph paper (e.g., draw 10 points on the graph of the equation $y = 3x$ and connect them by using a straight line).

Draw the points corresponding to linear relationships on graph paper.

1 Which ordered pair will make the equation $y = x + 2$ true?

A (2,0)
B (3,1)
C (0,2)
D (4,2)

SOLUTION/EXPLANATION

To find an ordered pair that solves this equation, replace x with the first number and y with the second number of each ordered pair. Check to see which equation is then true.

(2,0)	(3,1)	(0,2)
$y = x + 2$	$y = x + 2$	$y = x + 2$
$0 = 2 + 2$	$1 = 3 + 2$	$2 = 0 + 2$
$0 = 4$	$1 = 5$	$2 = 2$
not true	not true	true

The correct choice is: Ⓐ Ⓑ **Ⓒ** Ⓓ

2 When $x = 6$ in the equation $y = 2x + 7$, what ordered pair will you graph for the point?

A (6,5)
B (6,14)
C (6,19)
D (42,6)

SOLUTION/EXPLANATION

To find the ordered pair when the value of x is 6, replace x with 6 in the equation and solve.

$y = 2x + 7$
$y = 2 \times 6 + 7$
$y = 12 + 7$
$y = 19$

The ordered pair for this point is (6,19).

The correct choice is: Ⓐ Ⓑ **Ⓒ** Ⓓ

THINK ALONG

MG 2.2

Understand that the length of a horizontal line segment equals the difference of the *x*-coordinates.

Understand that the length of a horizontal line segment equals the difference of the *x*-coordinates.

1 What is the length of this segment?

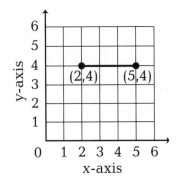

- **A** 1 unit
- **B** 2 units
- **C** 3 units
- **D** 4 units

SOLUTION/EXPLANATION

To find the length of a horizontal segment on a coordinate grid, count the units or subtract the *x*-coordinates.

(2,4) and (5,4) → 5 − 2 = 3

The length of this segment is 3 units.

The correct choice is: Ⓐ Ⓑ ⬤ Ⓓ

2 What is the length of this line segment?

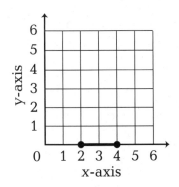

- **A** 1 unit
- **B** 2 units
- **C** 3 units
- **D** 4 units

SOLUTION/EXPLANATION

To find the length of a horizontal line segment in the coordinate plane, count the units or subtract the *x*-coordinates.

(2,0) and (4,0) → 4 − 2 = 2

The length of the segment is 2 units.

The correct choice is: Ⓐ ⬤ Ⓒ Ⓓ

Benchmark Assessment

MG 2.3

Understand that the length of a vertical line segment equals the difference of the y-coordinates.

Understand that the length of a vertical line segment equals the difference of the y-coordinates.

1 What is the length of this line segment?

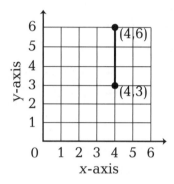

A 2 units
B 3 units
C 4 units
D 5 units

SOLUTION/EXPLANATION

To find the length of a vertical line segment on a coordinate grid, count the units or subtract the y-coordinates.

(4,6) and (4,3) → 6 − 3 = 3

The length of this segment is 3 units.

The correct choice is: Ⓐ **Ⓑ** Ⓒ Ⓓ

2 What is the length of this line segment?

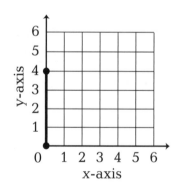

A 4 units
B 3 units
C 2 units
D 1 unit

SOLUTION/EXPLANATION

To find the length of a vertical line segment in the coordinate plane, count the units or subtract the y-coordinates.

(0,4) and (0,0) → 4 − 0 = 4

The length of the segment is 4 units.

The correct choice is: **Ⓐ** Ⓑ Ⓒ Ⓓ

T-22

Benchmark Assessment

Name _____

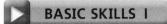

ADDITION FACTS TEST

	M	N	O	P
A	5 + 4 = 9	4 + 6 = 10	5 + 7 = 12	9 + 7 = 16
B	6 + 5 = 11	4 + 7 = 11	6 + 7 = 13	4 + 5 = 9
C	8 + 6 = 14	7 + 4 = 11	5 + 3 = 8	7 + 8 = 15
D	7 + 3 = 10	3 + 6 = 9	8 + 7 = 15	4 + 8 = 12
E	5 + 9 = 14	8 + 9 = 17	9 + 9 = 18	5 + 5 = 10
F	5 + 8 = 13	9 + 8 = 17	9 + 3 = 12	3 + 9 = 12
G	3 + 7 = 10	3 + 4 = 7	4 + 9 = 13	6 + 4 = 10
H	9 + 4 = 13	3 + 8 = 11	7 + 6 = 13	8 + 4 = 12
I	6 + 8 = 14	9 + 5 = 14	7 + 9 = 16	9 + 6 = 15
J	3 + 3 = 6	5 + 6 = 11	3 + 5 = 8	8 + 8 = 16
K	4 + 3 = 7	7 + 5 = 12	6 + 3 = 9	4 + 4 = 8
L	8 + 5 = 13	8 + 3 = 11	6 + 9 = 15	7 + 7 = 14

Benchmark Assessment

SUBTRACTION FACTS TEST

	M	N	O	P
A	16 − 7 = 9	11 − 5 = 6	12 − 6 = 6	11 − 6 = 5
B	15 − 7 = 8	17 − 9 = 8	9 − 6 = 3	12 − 4 = 8
C	11 − 8 = 3	13 − 9 = 4	14 − 6 = 8	10 − 6 = 4
D	17 − 8 = 9	14 − 9 = 5	14 − 5 = 9	10 − 4 = 6
E	10 − 7 = 3	12 − 7 = 5	11 − 3 = 8	15 − 9 = 6
F	12 − 5 = 7	8 − 3 = 5	11 − 4 = 7	14 − 7 = 7
G	16 − 9 = 7	11 − 7 = 4	6 − 3 = 3	9 − 3 = 6
H	7 − 4 = 3	14 − 8 = 6	10 − 3 = 7	13 − 4 = 9
I	12 − 9 = 3	16 − 8 = 8	15 − 8 = 7	12 − 3 = 9
J	8 − 5 = 3	9 − 5 = 4	15 − 6 = 9	9 − 4 = 5
K	8 − 4 = 4	13 − 5 = 8	10 − 5 = 5	13 − 7 = 6
L	12 − 8 = 4	13 − 8 = 5	18 − 9 = 9	7 − 3 = 4

Benchmark Assessment

Name _____

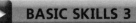

ADDITION FACTS TEST

	M	N	O	P	Q	R	S	T
A	3 + 5	7 + 5	2 + 2	3 + 4	7 + 3	9 + 6	4 + 8	9 + 4
B	7 + 4	2 + 7	8 + 2	5 + 1	2 + 4	7 + 7	9 + 3	2 + 5
C	6 + 3	6 + 9	6 + 8	9 + 5	9 + 7	2 + 8	4 + 7	2 + 3
D	7 + 6	6 + 4	8 + 4	1 + 4	1 + 9	4 + 5	7 + 2	8 + 1
E	1 + 5	9 + 1	8 + 3	9 + 8	5 + 8	2 + 6	3 + 9	4 + 9
F	9 + 2	4 + 1	6 + 7	5 + 6	9 + 9	6 + 6	3 + 1	7 + 1
G	8 + 5	1 + 7	5 + 4	5 + 3	1 + 8	3 + 7	3 + 6	8 + 8
H	5 + 5	1 + 2	1 + 3	3 + 2	3 + 3	4 + 2	4 + 4	1 + 1
I	7 + 9	5 + 7	2 + 1	8 + 6	6 + 2	6 + 5	7 + 8	4 + 3
J	5 + 2	4 + 6	8 + 9	5 + 9	6 + 1	2 + 9	3 + 8	1 + 6

Benchmark Assessment

SUBTRACTION FACTS TEST

	M	N	O	P	Q	R	S	T
A	12 − 9	3 − 2	11 − 7	8 − 1	14 − 9	7 − 4	11 − 2	8 − 7
B	8 − 6	6 − 3	10 − 1	8 − 4	14 − 7	9 − 1	9 − 6	6 − 5
C	14 − 6	5 − 2	11 − 9	11 − 3	4 − 2	10 − 2	8 − 2	9 − 4
D	15 − 8	13 − 7	9 − 2	17 − 8	7 − 5	11 − 8	11 − 6	8 − 5
E	9 − 5	12 − 5	12 − 6	9 − 3	12 − 7	10 − 6	8 − 3	6 − 2
F	5 − 4	17 − 9	4 − 1	16 − 8	10 − 9	10 − 3	10 − 4	14 − 5
G	10 − 8	15 − 9	11 − 5	15 − 6	7 − 2	6 − 1	13 − 6	13 − 4
H	12 − 8	14 − 8	12 − 4	5 − 1	7 − 1	16 − 9	7 − 3	13 − 8
I	5 − 3	13 − 9	9 − 7	10 − 5	2 − 1	6 − 4	11 − 4	9 − 8
J	16 − 7	10 − 7	18 − 9	15 − 7	7 − 6	13 − 5	3 − 1	12 − 3

Benchmark Assessment

MULTIPLICATION FACTS TEST

	M	N	O	P
A	10 × 6 = 60	9 × 11 = 99	12 × 3 = 36	5 × 5 = 25
B	10 × 3 = 30	5 × 8 = 40	1 × 4 = 4	10 × 8 = 80
C	1 × 9 = 9	3 × 8 = 24	1 × 1 = 1	1 × 5 = 5
D	10 × 9 = 90	5 × 10 = 50	9 × 10 = 90	12 × 1 = 12
E	5 × 11 = 55	3 × 9 = 27	8 × 3 = 24	10 × 10 = 100
F	4 × 8 = 32	3 × 1 = 3	11 × 7 = 77	11 × 9 = 99
G	8 × 10 = 80	3 × 6 = 18	5 × 3 = 15	5 × 7 = 35
H	2 × 5 = 10	10 × 1 = 10	4 × 6 = 24	8 × 6 = 48
I	7 × 8 = 56	7 × 1 = 7	7 × 7 = 49	11 × 4 = 44
J	10 × 4 = 40	6 × 10 = 60	6 × 6 = 36	9 × 8 = 72
K	12 × 9 = 108	1 × 7 = 7	3 × 7 = 21	1 × 12 = 12
L	5 × 9 = 45	4 × 10 = 40	2 × 11 = 22	4 × 5 = 20

Benchmark Assessment

DIVISION FACTS TEST

BASIC SKILLS 6

	M	N	O	P
A	36 ÷ 6 = 6	12 ÷ 12 = 1	48 ÷ 4 = 12	48 ÷ 6 = 8
B	90 ÷ 9 = 10	42 ÷ 6 = 7	40 ÷ 10 = 4	56 ÷ 7 = 8
C	108 ÷ 9 = 12	9 ÷ 1 = 9	21 ÷ 3 = 7	90 ÷ 10 = 9
D	60 ÷ 12 = 5	16 ÷ 8 = 2	30 ÷ 3 = 10	36 ÷ 3 = 12
E	20 ÷ 10 = 2	27 ÷ 3 = 9	20 ÷ 5 = 4	18 ÷ 2 = 9
F	42 ÷ 7 = 6	120 ÷ 12 = 10	45 ÷ 9 = 5	120 ÷ 10 = 12
G	100 ÷ 10 = 10	18 ÷ 6 = 3	22 ÷ 11 = 2	27 ÷ 9 = 3
H	30 ÷ 5 = 6	3 ÷ 3 = 1	66 ÷ 6 = 11	64 ÷ 8 = 8
I	10 ÷ 5 = 2	18 ÷ 3 = 6	121 ÷ 11 = 11	32 ÷ 8 = 4
J	8 ÷ 2 = 4	4 ÷ 2 = 2	144 ÷ 12 = 12	35 ÷ 7 = 5
K	108 ÷ 12 = 9	66 ÷ 11 = 6	10 ÷ 1 = 10	99 ÷ 9 = 11
L	96 ÷ 8 = 12	60 ÷ 6 = 10	6 ÷ 1 = 6	7 ÷ 7 = 1

Benchmark Assessment

Name _____

BASIC SKILLS 7

MULTIPLICATION FACTS TEST

	M	N	O	P	Q	R	S	T
A	4 × 4	5 × 9	8 × 6	11 × 6	5 × 5	10 × 12	5 × 6	4 × 5
B	7 × 9	7 × 12	9 × 11	10 × 5	11 × 4	12 × 9	12 × 5	9 × 5
C	11 × 12	9 × 6	4 × 12	10 × 7	10 × 4	12 × 8	10 × 6	7 × 8
D	8 × 10	10 × 11	6 × 10	10 × 8	4 × 8	4 × 8	8 × 5	6 × 11
E	10 × 9	8 × 7	7 × 6	8 × 11	4 × 11	9 × 4	6 × 5	12 × 6
F	9 × 7	5 × 11	12 × 4	12 × 12	11 × 5	5 × 8	5 × 7	4 × 10
G	11 × 9	7 × 7	7 × 5	12 × 7	5 × 4	8 × 12	8 × 9	4 × 9
H	7 × 4	11 × 10	6 × 9	6 × 8	7 × 10	9 × 8	6 × 6	12 × 10
I	11 × 7	8 × 8	10 × 10	11 × 11	6 × 7	9 × 12	12 × 11	6 × 4
J	6 × 12	7 × 11	9 × 9	5 × 12	5 × 10	9 × 10	4 × 6	11 × 8

Benchmark Assessment

Name _____

BASIC SKILLS 8

DIVISION FACTS TEST

	M	N	O	P	Q	R	S	T
A	5)60 = 12	4)48 = 12	5)20 = 4	11)44 = 4	5)30 = 6	10)90 = 9	10)70 = 7	8)48 = 6
B	4)44 = 11	11)110 = 10	10)120 = 12	11)121 = 11	10)100 = 10	12)132 = 11	9)54 = 6	11)88 = 8
C	12)84 = 7	5)55 = 11	7)56 = 8	7)28 = 4	5)35 = 7	9)108 = 12	6)36 = 6	7)84 = 12
D	8)56 = 7	7)70 = 10	8)88 = 11	8)80 = 10	12)72 = 6	6)66 = 11	9)90 = 10	5)25 = 5
E	12)48 = 4	5)40 = 8	10)50 = 5	9)81 = 9	7)77 = 11	8)96 = 12	12)120 = 10	9)63 = 7
F	5)45 = 9	6)48 = 8	4)32 = 8	7)63 = 9	6)72 = 12	8)64 = 8	4)36 = 9	11)55 = 5
G	10)40 = 4	11)132 = 12	6)60 = 10	12)144 = 12	6)30 = 5	4)40 = 10	9)99 = 11	7)49 = 7
H	12)108 = 9	9)45 = 5	8)72 = 9	4)20 = 5	6)42 = 7	8)32 = 4	6)54 = 9	4)16 = 4
I	4)24 = 6	7)35 = 5	8)40 = 5	4)28 = 7	12)60 = 5	6)24 = 4	9)36 = 4	11)77 = 7
J	7)42 = 6	10)60 = 6	10)110 = 11	10)80 = 8	11)99 = 9	9)72 = 8	5)50 = 10	12)96 = 8

Benchmark Assessment

Name _____

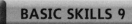

2-DIGIT ADDITION

	M	N	O	P
A	10 + 64 = 74	25 + 60 = 85	50 + 25 = 75	59 + 20 = 79
B	10 + 63 = 73	17 + 50 = 67	50 + 43 = 93	50 + 40 = 90
C	40 + 59 = 99	73 + 10 = 83	30 + 37 = 67	15 + 40 = 55
D	10 + 43 = 53	24 + 20 = 44	30 + 67 = 97	16 + 50 = 66
E	20 + 56 = 76	19 + 80 = 99	40 + 28 = 68	67 + 10 = 77
F	70 + 25 = 95	23 + 50 = 73	30 + 18 = 48	16 + 80 = 96
G	60 + 35 = 95	52 + 10 = 62	10 + 61 = 71	46 + 50 = 96
H	70 + 16 = 86	16 + 30 = 46	20 + 58 = 78	44 + 30 = 74
I	30 + 25 = 55	69 + 10 = 79	40 + 44 = 84	33 + 50 = 83
J	50 + 35 = 85	32 + 30 = 62	20 + 17 = 37	71 + 20 = 91
K	50 + 26 = 76	29 + 60 = 89	10 + 29 = 39	30 + 60 = 90
L	20 + 65 = 85	59 + 10 = 69	40 + 38 = 78	12 + 50 = 62

Benchmark Assessment

2-DIGIT SUBTRACTION

	M	N	O	P	Q	R	S	T
A	94 − 84	52 − 20	64 − 44	80 − 30	84 − 74	87 − 50	35 − 25	21 − 10
B	82 − 42	91 − 30	98 − 18	52 − 30	42 − 32	96 − 40	82 − 62	75 − 50
C	47 − 37	67 − 20	28 − 18	42 − 20	95 − 25	83 − 30	74 − 34	56 − 40
D	95 − 85	83 − 40	87 − 27	25 − 10	91 − 71	82 − 60	70 − 20	31 − 20
E	47 − 27	85 − 60	97 − 87	89 − 50	99 − 89	50 − 20	33 − 13	48 − 20
F	79 − 69	58 − 40	73 − 53	77 − 20	94 − 44	80 − 60	22 − 12	88 − 40
G	55 − 15	29 − 10	82 − 12	70 − 20	92 − 22	70 − 10	98 − 58	40 − 10
H	44 − 14	56 − 20	66 − 26	23 − 10	41 − 11	66 − 50	68 − 58	51 − 30
I	40 − 20	63 − 40	93 − 23	97 − 20	53 − 23	85 − 50	99 − 79	75 − 10
J	96 − 26	82 − 10	68 − 48	38 − 10	88 − 58	84 − 50	76 − 26	89 − 60

Benchmark Assessment

Name _____

NUMBER SENSE

NS 1.0

Write the value of the bold digit.

1. 9,3**6**1 _____
2. **3**8,615 _____
3. **4**73,892 _____
4. 2,5**7**6,497 _____
5. **1**,416,975 _____
6. 81,98**2**,510 _____

NS 1.1

Write each number in two other forms.

7. 40,000 + 700 + 4 _____
_____

8. 5,810,000 _____
_____

9. seven hundred sixty thousand, nine hundred fifty _____
_____

10. 3,000,000 + 60,000 + 90 + 3 _____
_____

NS 1.2

Compare. Write <, >, or = in each ◯.

11. 14,837 ◯ 14,378
12. 54,801 ◯ 5,987
13. 295,649 ◯ 298,102
14. 2,376,780 ◯ 2,095,689
15. 487,901 ◯ 487,901
16. 379,654 ◯ 3,591,005

Benchmark Assessment

Name _____

NUMBER SENSE

NS 1.3

Round each number to the place value of the bold digit.

17. 2**1**6,307   18. 872,**1**60   19. 3,68**2**,147
    200,000   872,200   3,682,000

20. 4,**7**35,086   21. **9**,052,178   22. 8,**4**51,658
    4,740,000   9,000,000   8,500,000

NS 1.4

23. The total attendance at the state fair was 237,691. Round 237,691 to the nearest hundred thousand, then ten thousand, and thousand. Which of these rounded amounts is closest to the actual attendance? Explain.

    200,000; 240,000; 238,000;

    Rounding to the lesser place

    value gives you a number

    closest to the actual number.

24. The newspaper reported that about 5,000 people attended the summer concert. Which of these numbers could have been the actual attendance if the newspaper rounded to the nearest thousand?

    A 5,812
    B 5,597
    Ⓒ 4,506
    D 4,499

NS 2.1

Estimate the sum or difference. Possible estimates are given.

25. 78,641   26. 32,905   27. 67,478   28. $59,154
    +12,652   −11,406   −55,695   +$27,235
     90,000   20,000   10,000 $90,000

29. $74,564   30. 19,390   31. 795,871   32. $854,210
    −$29,487   +34,517   −439,036   +$128,409
     $40,000   50,000   400,000 $1,000,000

12   Benchmark Assessment

Name _____

> **NUMBER SENSE**

NS 3.0

33. An orchard puts 25 apples in each bag it sells. On Saturday morning 231 bags were packed. At the end of the day, 48 bags were not sold. How many apples were sold?

   A 183
   B 1,200
   C 1,281
   (D) 4,575

34. The library owns 543 CDs. Of these, 348 have been checked out. How many CDs are still in the library?

   A 105
   (B) 195
   C 205
   D 891

NS 3.1

Find the sum or difference.

35. 4,852
    +2,546

     7,398

36. 6,915
    −3,745

     3,170

37. 9,103
    +5,787

    14,890

38. 7,254
    −4,836

     2,418

39. 5,219
    −4,628

       591

40. 12,018
    +31,995

     44,013

41. 898,127
    +426,594

    1,324,721

42. 516,418
    −122,397

     394,021

NS 4.0

Find the product.

43. 4 × 12   44. 9 × 8   45. 8 × 11
      48   72   88

46. 8 × 7   47. 6 × 12   48. 9 × 7
      56   72   63

Benchmark Assessment 13

Name _____

ALGEBRA AND FUNCTIONS

AF 1.0

Find the value of the variable.
Write a related equation. One possible equation is given.

1. $15 \div 5 = x$
   $x = 3$
   $3 \times 5 = 15$

2. $3 \times 6 = n$
   $n = 18$
   $18 \div 3 = 6$

3. $a \div 4 = 4$
   $a = 16$
   $4 \times 4 = 16$

4. $8 \times d = 32$
   $d = 4$
   $32 \div 4 = 8$

AF 1.1

Write an expression that matches the words.

5. a number of CDs, n, plus 5 CDs

   $n + 5$

6. 17 pretzels minus a number of pretzels, x

   $17 - x$

7. a number of pencils, p, divided by 8 students

   $p \div 8$

8. $56 divided by a number of friends, f

   $\$56 \div f$

9. Kelly picked 14 tulips. Kate gave her some more. Then Kelly had 21 tulips. Which equation describes this?
   A $14 + 21 = n$
   B $n - 21 = 14$
   C $21 + n = 14$
   D $14 + n = 21$

10. The tackle shop rented 9 fishing rods. By noon only 3 fishing rods were left. Which equation describes this?
    A $9 - 3 = n$
    B $n - 9 = 3$
    C $n + 3 = 9$
    D $9 - n = 3$

AF 1.2

Find the value of each expression.

11. $254 + (75 - 36)$

    293

12. $(125 - 87) + 115$

    153

13. $4{,}065 - (372 + 283)$

    3,410

14   Benchmark Assessment

Name _____

ALGEBRA AND FUNCTIONS

AF 1.3

Write the expression and solve.

14. At an art exhibit, there were 54 paintings and 75 photographs. The artist sold 48 photographs. How many pieces of art were left?

    _____54 + (75 − 48);_____

    _____81 pieces of art_____

15. At the fruit stand, there were 47 watermelons and some honeydew melons. The fruit stand sold 25 watermelons. There were 33 melons left. How many honeydews were there?

    _____$n + (47 − 25)$_____

    _____$n = 11$ honeydew_____

AF 1.5

Find a rule. Write the rule as an equation. Possible rules are given.

16.
Input	Output
x	y
3	18
5	30
7	42

_____$y = 6 \times x$_____

17.
Input	Output
a	b
35	7
40	8
45	9

_____$b = a \div 5$_____

18.
Input	Output
c	d
6	48
8	64
10	80

_____$d = 8 \times c$_____

Which equation describes the rules in these tables?

19.
Input x	33	43	50	68
Output y	27	37	46	62

A $x + 10 = y$
B $x − 10 = y$
C $6 \times x = y$
D $x − 6 = y$

20.
Input p	13	17	21	24
Output r	26	30	34	37

A $2 \times p = r$
B $p + 13 = r$
C $p + 4 = r$
D $p − 13 = r$

Benchmark Assessment

Name _____

ALGEBRA AND FUNCTIONS

AF 2.0

Complete to make the equation true.

21. 12 + 5 = 5 + ☐   22. 20 + ☐ = 10 + 8 + 8   23. 7 + 9 = ☐ + 3 + 5

   _____12_____   _____6_____   _____8_____

Tell whether the values on both sides of the equation are equal. Explain.

24. 1 quarter + 2 nickels = 3 dimes + 5 pennies

_____yes; 35¢ = 35¢_____

25. 2 quarters + 4 dimes + 1 nickel + 4 pennies = 10 dimes and 4 pennies

_____no; 99¢ does not equal $1.04_____

AF 2.1

Complete to make the equation true. *Possible answers are given.*

26. 2 dimes + 1 nickel = 4 nickels + ___5 pennies_____

27. 1 nickel + 4 dimes + 3 pennies = 1 quarter + 2 dimes + ___3 pennies___

28. 1 quarter + 3 dimes + 2 nickels = 2 quarters + 1 dime + ___1 nickel___

Choose the value that completes the equation.

29. 1 quarter + 3 nickels = 3 dimes + ☐

    A 1 nickel   C 1 quarter
    (B) 2 nickels   D 5 pennies

30. 5 dimes + ☐ + 3 pennies = 3 quarters + 3 pennies

    (A) 1 quarter   C 10 pennies
    B 2 nickels   D 3 dimes

AF 2.2

Multiply both sides by the given number. Find the new value.

31. 8 = (2 × 4)
    Multiply both sides by 3.

    _____24 = 24_____

32. (3 × 4) = (10 + 2)
    Multiply both sides by 12.

    _____144 = 144_____

Name _____

STATISTICS, DATA ANALYSIS, AND PROBABILITY

SDAP 1.0

For Problems 1–3, use the frequency table.

MILES TRAVELED		
Week	Frequency	Cumulative Frequency
1	456	456
2	623	1,079
3	347	1,426
4	595	2,021

1. Complete the frequency table.
2. How many miles were driven in all? __2,021__
3. Explain how you found the frequency for the fourth week.

   Possible answer: I added 595 to the sum of the miles from the three previous weeks.

SDAP 1.1

For Problems 4–6, use the line plot.

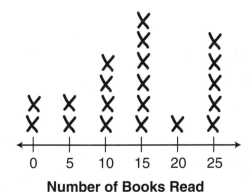

Number of Books Read

4. Students were asked how many books they read last year. How many students read more than 10 books?

   A 9
   B 10
   C 11
   D 12

5. How many students were surveyed?

   A 0
   B 6
   C 20
   D 25

6. How many books did most students read?

   A 5
   B 10
   C 15
   D 20

Benchmark Assessment 17

Name _____

STATISTICS, DATA ANALYSIS, AND PROBABILITY

SDAP 1.2

For Problems 7–8, use the table.

SARA'S TEST SCORES							
Test	1	2	3	4	5	6	7
Score	92	87	90	87	95	93	85

7. What is the median of Sara's test scores? Explain how you found the median.

   90; I ordered the scores from least to greatest and found the

   middle number.

8. What is the mode of Sara's test scores? Explain how you found the mode.

   87; 87 is the score that occurs most often.

SDAP 1.3

For Problems 9–10, use the bar graph.

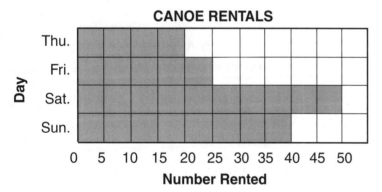

9. How many more canoes were rented on Saturday than on Thursday?

   A 20
   B 25
   C) 30
   D 40

10. How many canoes were rented on all four days?

    A 45
    B 95
    C 135
    D 145

18 Benchmark Assessment

Name _____

> **MATHEMATICAL REASONING**

MR 1.0

For Problems 1–2, use the graph.

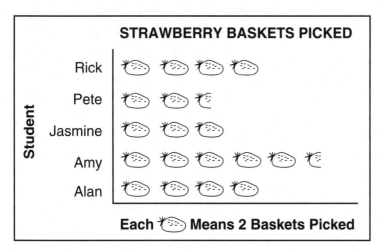

1. How many baskets did Alan and Jasmine pick together?

   A 7
   B 8
   C 14
   D 19

2. How many more baskets did Amy pick than Pete?

   A 5
   B 6
   C 9
   D 11

MR 1.1

3. Use the table to make a schedule for Mike between 10:00 A.M. and 3:30 P.M.

THINGS TO DO		SCHEDULE
Activity	**Time**	Possible Schedule
Skating	45 min	10:00–10:45 A.M. Skating
Lunch	30 min	11:00–11:30 A.M. Haircut
Homework	1 hr	11:30–12:00 P.M. Lunch
Movie	12:10–2:20 P.M.	12:10–2:20 P.M. Movie
Haircut	11:00–11:30 A.M.	2:30–3:30 P.M. Homework

Benchmark Assessment

Name _____

MATHEMATICAL REASONING

MR 2.0

4. Paulo, Jill, Kim and Brian each hold a card. Paulo holds an even number. Jill's number is divisible by 3 but not by 9. Kim's number is a multiple of 10. Who holds number 27?

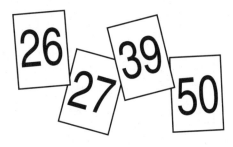

- A Paulo
- (B) Brian
- C Jill
- D Kim

5. Three children bought ice cream. Each bought a different flavor. Mayella always gets vanilla. Sara did not choose strawberry. Joe never had chocolate. Which of the following is a reasonable conclusion?
   - A Sara chose strawberry ice cream.
   - (B) Joe chose strawberry ice cream.
   - C Mayella chose strawberry ice cream.
   - D Sara chose vanilla ice cream.

MR 2.1

6. Harry sells magazine subscriptions. He sold $486 in June, $352 in July, and $258 in August. What are his total sales for the three months?

   Estimate. Then decide which answer is reasonable.

   - A $196
   - B $996
   - (C) $1,096
   - D $10,096

7. Juan has $0.76. Georgio has $0.67. Mary has $0.66. Brooke has $0.70. How much money do they have in all?

   Estimate. Then decide which answer is reasonable.

   - A $0.79
   - B $1.79
   - (C) $2.79
   - D $3.79

8. Tom is 89 years old. Joe is 32. Neil said Tom is 47 years older than Joe. Was his answer reasonable? If not, estimate a reasonable answer.

   No; Possible answer: 89 rounds to 90 and 32 rounds to 30.

   A reasonable answer is 60 because 90 − 30 = 60.

20   **Benchmark Assessment**

Name _____

MATHEMATICAL REASONING

MR 2.3

For Problems 9–10, use the table.

TICKETS SOLD		
Day	Fourth Grade Students	Fifth Grade Students
1	14	11
2	12	9
3	10	12
4	11	7

9. Make a double-bar graph to compare the data of the Grade 4 and Grade 5 classes. Possible graph is shown.

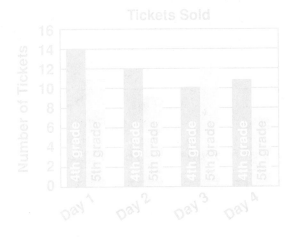

10. Can you conclude that the fourth grade students sold more tickets than the fifth grade students? Explain.

Possible answer: Yes; on three out of the four days, they sold more tickets and on the fourth day they only sold two tickets fewer than the fifth graders.

MR 3.1

11. Mona used her calculator to find 882 − 561. It showed an answer of 121, but she is not sure she entered the numbers correctly. Is that answer reasonable? If not, estimate a reasonable answer.

No; a reasonable estimate is 900 − 600 = 300.

12. Oscar has 290 CDs. He has only 103 CD boxes. Oscar says he needs 393 more boxes to have a box for each CD. Is his answer reasonable? If not, estimate a reasonable answer.

No; a reasonable estimate is 300 − 100 = 200.

Benchmark Assessment   21

Name _____

MATHEMATICAL REASONING

MR 2.6

13. The number of students enrolled in Brown School is 1,628. The number of students enrolled in Hart School is 1,546, and the number of students enrolled in Revere School is 1,829. How many more students are enrolled in Revere School than in Hart School?

    _____283 students_____

14. Mike has 541 photographs. Beth has 762 photographs. Dan has 536 photographs, and Tina has 458 photographs. How many more photographs does Tina need in order to have as many as Beth?

    _____304 photographs_____

For Problems 15–16, use the table to solve.

AMUSEMENT PARK ATTENDANCE	
April	32,098
May	38,824
June	65,013
July	68,944
August	76,025
September	29,739

15. The sum of attendance figures of the two lowest months is less than the attendance of which other months?

    A August
    B July, August
    C June, July, August
    D May, June, July, August

16. What is the sum of the attendance of the two busiest months?

    A 145,969
    B 144,969
    C 141,038
    D 133,957

Benchmark Assessment

Name _____

UNITS 1–3 • PAGE 1

Write the correct answer.

1. **NS 1.1** Write the value of the digit 5 in the number 52,981.

   _____50,000_____

2. **NS 1.1** What is three hundred twenty-two thousand, one hundred fifteen in standard form?

   _____322,115_____

3. **NS 1.3** What is 1,402,945 rounded to the nearest thousand?

   _____1,403,000_____

4. **NS 1.3** What is 1,243 + 4,907 rounded to the nearest hundred?

   _____6,100_____

Choose the letter of the correct answer.

5. **NS 1.0** In which place is the digit 2 in the number 203,987?

   A ten thousands
   B thousands
   C hundreds
   D Not Given

6. **NS 1.1** What is two million, six hundred nine thousand, four hundred sixty written in standard form?

   A 2,690,460
   B 2,609,460
   C 2,069,460
   D 2,069,406

7. **NS 1.2** Which shows the numbers ordered from **least** to **greatest**?

   A 4,678 4,768 4,867 4,876
   B 4,876 4,867 4,768 4,678
   C 4,678 4,768 4,876 4,867
   D Not Given

8. **NS 1.2** Which shows the numbers ordered from **greatest** to **least**?

   A 583,291 583,309 583,311
   B 583,311 583,291 583,309
   C 583,311 583,309 583,291
   D 583,309 583,311 583,291

Benchmark Assessment

23

Name _____

UNITS 1–3 • PAGE 2

Write the correct answer.

9. **NS 1.4** It takes Roberta one hour and 30 minutes to do her homework, 20 minutes to clean her room, and 15 minutes to walk the dog. Her friends will be over in 2 hours. Will she be finished before they come over? Explain.

No; she will finish 5 minutes after they arrive.

10. **NS 2.1** There are 301 students who attend Vine School. There are 217 students who attend Lake School. What is a reasonable estimate of the total number of students who attend both schools? Explain.

Possible answer: 500 students; 301 rounds to 300, and 217 rounds to 200; 300 + 200 = 500

Choose the letter of the correct answer.

11. **NS 1.4** Marvin wants to buy a hot dog for $1.75, French fries for $1.50, and a soda for $0.95. How much money will he need to buy the food?

   A $2
   B $3
   C $4
   D $5

12. **NS 2.1** There were 418 people at the museum on Tuesday and 386 people on Wednesday. About how many people attended the museum on both days?

   A 700
   B 750
   C 800
   D 900

13. **NS 3.1**   4,789
              + 2,123

   A 6,802
   B 6,812
   C 6,912
   D Not Given

14. **NS 3.1**   50,895
              − 42,786

   A 18,119
   B 18,109
   C 8,119
   D 8,109

15. **NS 1.4** **Write About It** Explain how you can use rounding to help you solve Problem 11.

$1.75 rounds to $2.00, $1.50 rounds to $2.00, and $0.95 rounds to $1.00. $2.00 + $2.00 + $1.00 = $5.00. So, Marvin will need about $5.00 to buy the food.

Benchmark Assessment

Name _____

UNITS 1–3 • PAGE 3

Write the correct answer.

16. **NS 3.0** Juanita noticed that the store had 11 cartons of eggs left. Each carton contained 12 eggs. How many eggs did the store have left?

_____ 132 eggs _____

17. **NS 2.1** Cathy collected 80 stamps. She put 5 stamps on each page in her scrapbook. How many pages did she use?

_____ 16 pages _____

18. **AF 1.0** $3 \times w = 36$. What is the value of w?

_____ $w = 12$ _____

19. **AF 1.0** What is the value of n? $n \div 6 = 6$

_____ $n = 36$ _____

Choose the letter of the correct answer.

20. **AF 1.0** Which two smaller arrays can be used to find the product 5×9?

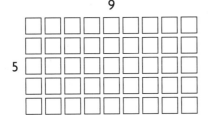

A 5×6 and 5×3
B 5×6 and 5×4
C 5×5 and 5×2
D 5×3 and 5×3

21. **AF 1.0** Find the product. Use the array.

$5 \times 12 = \Box$

A 50
B 60
C 65
D 72

Benchmark Assessment

Name _____

UNITS 1–3 • PAGE 4

Write the correct answer.

22. **AF 1.1** Jan is putting 42 tomato plants in rows in her garden. She puts 6 tomato plants in each row. Write an equation she could use to find the total number of rows of tomato plants, r, in her garden.

    $42 \div 6 = r$, or $6 \times r = 42$

23. **AF 1.1** Jenny is going to bake 12 tins of muffins. Each tin holds 9 muffins. Write an equation that could be used to find the total number of muffins, m, that Jenny will bake.

    $9 \times 12 = m$

24. **AF 1.5** The rule for an input-output table is $y = x \div 4$. If $y = 12$, what is x?

    $x = 48$

25. **AF 2.0** If $q = 38$, what value of p will make this equation true?

    $p + 18 = q$

    $p = 20$

Choose the letter of the correct answer.

26. **AF 1.2** What is the value of the expression $32 - (24 \div 4)$?

    A 2
    B 6
    C 26
    D Not Given

27. **AF 1.2** What is the value of the expression $9 + (3 \times 4)$?

    A 12
    B 16
    C 21
    D 48

28. **AF 1.5** The rule for an input-output table is $y = x - 8$. If $y = 10$, what is x?

    A $x = 18$
    B $x = 20$
    C $x = 24$
    D $x = 28$

29. **AF 2.0** Which equation is true?

    A $20 + 12 = 20 + 13$
    B $19 - 4 = 19 - 4 + 2$
    C $21 + 6 + 5 = 21 + 11$
    D $14 + 12 = 14 + 8 + 2$

Benchmark Assessment

Name _____

UNITS 1–5 • PAGE 5

Write the correct answer.

30. **AF 2.1** Jill has 2 dimes and 3 quarters. Bill has 2 quarters and 4 dimes. Do they have the same amount of money? Explain.

   No; Jill has $0.95
   and Bill has $0.90.

31. **AF 2.1** Colleen and Brian have to do chores for the same amount of time. On Tuesday, Colleen worked for 35 minutes, and on Wednesday for 25 minutes. Brian worked for 15 minutes on Tuesday and 15 minutes on Wednesday. How many more minutes does he have to work to be equal with Colleen? Explain.

   30 minutes; Colleen worked
   60 minutes. Brian worked only
   30 minutes; 60 − 30 = 30 min

32. **AF 2.2** Maria has 4 dimes. Kari has 8 nickels. If their money amounts are multiplied by 3, how much money will each girl have?

   $1.20

33. **AF 2.2** Multiply both sides by 6. What is the new value?

   $2 \times 4 = 5 + 3$

   48 = 48

Choose the letter of the correct answer.

34. **AF 2.2** Maddy makes 24 brownies for a bake sale. She leaves 6 at home for her family. Then she buys 12 more at the store. Which expression shows how many brownies she brings to the bake sale?

   A 24 − (6 + 12)
   B (24 − 6) + 12
   C (24 − 12) + 6
   D 24 + 6 + 12

35. **AF 2.2** Janine, Joelle, and Jackie each brought 12 brownies to the school fair. Before it started, they each ate one brownie. Which expression shows how many brownies they had left to sell?

   A 12 − (3 × 3)
   B (12 × 3) + 3
   C (12 × 3) − 3
   D 12 × (4 − 3)

Benchmark Assessment

Name _____

UNITS 1–3 • PAGE 6

Write the correct answer.

36. SDAP 1.0 How many toy cars and toy boats are in the Antique Toy Museum?

ANTIQUE TOY MUSEUM TOYS	
Train	156
Car	278
Airplane	216
Boat	197

_____475_____

37. SDAP 1.0 The school secretary wants to put this information into a bar graph. What is the most reasonable interval for her to use? Explain.

NUMBER OF ABSENCES AT REVERE SCHOOL	
Monday	12
Tuesday	23
Wednesday	8
Thursday	17
Friday	24

Possible answer: intervals of 5; The data goes up to 24. Intervals of 10 are too large and intervals of 2 are too small.

Choose the letter of the correct answer. For 38–41, use the line plot. The line plot shows the number of hours some students slept one night.

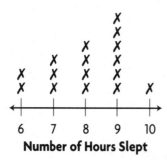
Number of Hours Slept

38. SDAP 1.2 What is the mode of the data?

A 7 hours   C 9 hours
B 8 hours   D 10 hours

39. SDAP 1.2 What is the median number of hours slept?

A 7 hours   C 9 hours
(B) 8 hours   D Not Given

40. SDAP 1.2 What is the range for the number of hours slept?

A 5   C 10
B 6   (D) Not Given

41. SDAP 1.3 How many students were surveyed?

A 9 students   (C) 16 students
B 10 students   D Not Given

Name _____

UNITS 1–3 • PAGE 7

Write the correct answer.

42. **SDAP 1.0** Which type of graph would be best for displaying these data?

TV FAVORITES	
Situation Comedy	36
Drama	22
Sports	16
Action/Adventure	18

bar graph

43. **SDAP 1.0** What is the difference between the amount of rainfall in March and in May?

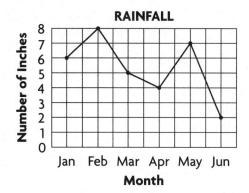

2 inches

Choose the letter of the correct answer.

44. **MR 2.0** In what order will the baseball teams play?

BASEBALL GAME STARTING TIMES	
Hawks	9:00 A.M.
Bears	11:00 A.M.
Rams	5:00 P.M.
Lions	3:00 P.M.

A Hawks, Bears, Lions, Rams
B Lions, Rams, Bears, Hawks
C Bears, Hawks, Lions, Rams
D Rams, Lions, Hawks, Bears

45. **MR 2.1** Laurel bought a shirt for $9.95, a pair of socks for $2.99, and a pair of jeans for $14.75. About how much did she spend in all?

Estimate. Then decide which answer is reasonable.

A $15
B $18
C $28
D $280

46. **MR 3.1** **Write About It** Explain how you know that your answer to Problem 45 is reasonable.

Possible answer: $9.95 rounds to $10.00, $2.99 rounds to $3.00, and $14.75 rounds to $15.00; $10.00 + $3.00 + $15.00 = $28.00

Benchmark Assessment

Name _____

UNITS 1–3 • PAGE 8

Write the correct answer.

47. **MR 1.0** If each activity lasts 30 minutes, what time will hiking end?

CAMP SCHEDULE	
Activity	Starting Time
Hiking	9:45 A.M.
Swimming	11:00 A.M.
Soccer	2:00 P.M.
Crafts	4:15 P.M.

10:15 A.M.

48. **MR 1.1** Tour bus rides through London start every half-hour. The first one is at 9 A.M. Merill wants to take a tour, but first she must spend 45 minutes having breakfast, and 20 minutes getting dressed. It is now 8:15 A.M. What time is the earliest tour bus Merill can catch?

9:30 A.M.

Choose the letter of the correct answer.

49. **MR 2.3** When was the greatest difference in goals scored between two games?

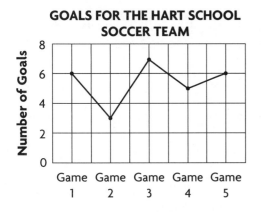

A between Games 1 and 2
B between Games 2 and 3
C between Games 3 and 4
D between Games 4 and 5

50. **MR 2.6** How much more allowance does Lea get than Kim?

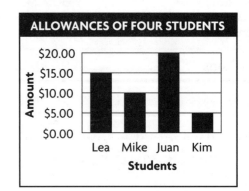

A $1.00
B $5.00
C $10.00
D $20.00

30 Benchmark Assessment

Name _____

NUMBER SENSE

NS 1.2

Compare. Write <, >, or = for each ◯.

1. 0.4 ⊚=⊚ 0.40
2. 0.68 ⊚<⊚ 0.8
3. 1.9 ⊚>⊚ 1.21

Order the decimals from *least* to *greatest*.

4. 3.6, 3.54, 3.82, 3.38, 2.4

   __2.4, 3.38, 3.54, 3.6, 3.82__

5. 1.7, 0.76, 1.17, 1.67, 1.76

   __0.76, 1.17, 1.67, 1.7, 1.76__

NS 1.4

Round one factor. Estimate the product. Possible estimates are given.

6. 238
   ×   8
   _____
   1,600

7. $6.77
   ×   4
   _____
   $28.00

8. $64
   ×   5
   _____
   $300

NS 1.5

Write the fraction or mixed number for the shaded part.

9.

   $\frac{2}{3}$

10.

    $3\frac{2}{3}$

11. Which is an equivalent fraction for $\frac{3}{4}$?

    A $\frac{3}{8}$
    B $\frac{5}{10}$
    C $\frac{6}{10}$
    D $\frac{9}{12}$

12. Which of the following fractions is written in simplest form?

    A $\frac{5}{6}$
    B $\frac{4}{6}$
    C $\frac{3}{9}$
    D $\frac{8}{12}$

Benchmark Assessment

Name _____

NUMBER SENSE

NS 1.6

Write each fraction as a decimal. Write each decimal as a fraction.

13. $\frac{7}{100}$ 14. $1\frac{1}{4}$ 15. 3.75   16. 0.73

     0.07   1.25 $3\frac{3}{4}$ $\frac{73}{100}$

17. Is a rope $3\frac{1}{4}$ feet long equivalent to a rope 3.4 feet long? Explain.

    No; one is 3.25 feet long, and the other is 3.40 feet long.

18. Megan eats $\frac{6}{10}$ of the pizza. Mike eats 0.4 of the pizza. Who eats more? Explain.

    Megan ate more than Mike. (0.6 > 0.4)

NS 1.7

Write the decimal and an equivalent fraction for each model.

19.

    1.3; $1\frac{3}{10}$

20.

    1.75; $1\frac{3}{4}$

NS 1.9

Use the number line to find the equivalent decimal and mixed number for the given letter.

21.

    A 1.3; $\frac{13}{10}$
    (B) 1.4; $\frac{14}{10}$
    C 1.5; $\frac{15}{10}$
    D 1.6; $\frac{16}{10}$

22.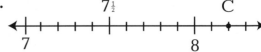

    A 7.9; $7\frac{9}{10}$
    B 8.0; 8
    C 8.1; $8\frac{1}{10}$
    (D) 8.2; $8\frac{2}{10}$

Name _____

NUMBER SENSE

NS 2.0

Find the sum or difference.

| 23. | 1.6
+ 0.9
2.5 | 24. | 8.83
+ 1.75
10.58 | 25. | 0.91
− 0.46
0.45 | 26. | $5.25
− $0.89
$4.36 |

NS 2.1

Estimate the sum or difference. Possible answers are given.

| 27. | 5.32
+ 8.78
14 | 28. | $2.09
+ $6.50
$9 | 29. | 85.17
− 13.76
80 | 30. | 45.80
− 12.99
40 |

| 31. | 36.75
+ 10.80
50 | 32. | 4.57
+ 6.15
11 | 33. | $27.00
− $ 8.42
$20 | 34. | $88.73
− $79.10
$10 |

NS 2.2

Round to the nearest tenth or ten cents.

| 35. 9.36 | 36. $56.82 | 37. 0.75 | 38. 1.67 |
| 9.4 | $56.80 | 0.8 | 1.7 |

39. Kira buys a terrarium for $32.95 and a chameleon for $3.50. She pays with two twenty-dollar bills. Is her change of $13.55 reasonable? Explain.

No; she spent over $30 so her change should be less than $10.

40. Ben has $100 in his savings account until he buys a game for $21.99 and a book for $6.25. He estimates he has $70 left. Is that reasonable? Explain.

Yes; he spent about $30 ($21.99 is about $20 and $6.25 is about $10), so $70 is reasonable.

Benchmark Assessment

Name _____

NUMBER SENSE

NS 3.0

Use a basic fact and patterns to write each product.

41. 4 × 600   42. 500 × 6   43. 7 × 700

   _2,400_   _3,000_   _4,900_

44. A science teacher orders 5 dozen magnifying glasses. She hands out 28 to one class and 25 to another. How many are left?

   A 19
   B 7
   C 6
   D 5

45. Elena makes 13 bracelets one day and 9 the next. She sells them for $4 each at a fair. How much does she earn?

   A $22
   B $48
   C $76
   D $88

Divide. You may wish to use basic facts or models.

46. 27 ÷ 4   47. 38 ÷ 9   48. 50 ÷ 8

   _6 r3_   _4 r2_   _6 r2_

49. How long does a package of 24 juice boxes last if 5 juice boxes are used each week?

   4 weeks and 4 days

50. How many juice boxes are in each group?

   _5_

Write the numbers you would use to estimate the quotient. Then estimate. _Possible answers are given._

51. 94 ÷ 32   52. 440 ÷ 78   53. 284 ÷ 47

   _90 ÷ 30; 3_   _400 ÷ 80; 5_   _300 ÷ 50; 6_

Solve. Name the operation or operations you used.

54. Sixty players have a jumbo jar of 500 licorice sticks to split equally at the end of a tournament. How many will each player get? Is there any licorice left over?

   8 sticks; 20 left over; division and subtraction

55. After Dan practices the piano for 30 minutes a day for 90 days, he will receive a new game player from his dad. How many hours of practice does this equal?

   45 hours; multiplication and division

Name _____

NUMBER SENSE

NS 3.2

Round each factor. Estimate the product.

56. 88   57. 495   58.   96   59. 343
   × 23   × 71   × 99   × 58

   1,800   35,000   10,000   18,000
   _____   _____   _____   _____

Divide mentally. Write the basic division fact and the quotient.

60. 630 ÷ 7   61. 360 ÷ 9

    63 ÷ 7 = 9; 90   36 ÷ 9 = 4; 40
    _____   _____

62. 150 ÷ 5   63. 120 ÷ 3

    15 ÷ 5 = 3; 30   12 ÷ 3 = 4; 40
    _____   _____

NS 3.3

64. Karl does 12 pull-ups a day for 28 days. How many does he do altogether?

   A 376
   B 346
   C) 336
   D 84

65. Marisa earns $0.75 a day for cleaning a birdcage. How much does she earn after 120 days?

   A $60.00
   B $75.50
   C $89.00
   D) $90.00

NS 3.4

Divide and check.

66. 77 ÷ 4   67. 80 ÷ 7   68. 438 ÷ 5   69. 678 ÷ 6

    19 r1   11 r3   87 r3   113
    _____   _____   _____   _____

Find the mean.

70. 21; 16; 19; 12   71. $3.92; $4.65; $5.02

        17                             $4.53
    _____   _____

Benchmark Assessment   35

Name _____

NUMBER SENSE

NS 4.0

Find the factors for each number.

72. 16

1, 2, 4, 8, 16

73. 20

1, 2, 4, 5, 10, 20

74. 27

1, 3, 9, 27

75. 36

1, 2, 3, 4, 6, 9, 12, 18, 36

76. Which equation is possible for an array of 12 tiles?

    A $2 \times (3 \times 4)$
    B $4 \times (2 \times 2)$
    C $(2 \times 3) \times 1$
    (D) $2 \times (2 \times 3)$

77. Which equation is possible for an array of 24 tiles?

    A $(2 \times 2) \times 5$
    (B) $(2 \times 2) \times (2 \times 3)$
    C $(4 \times 3) \times 3$
    D $(2 \times 2) \times 8$

NS 4.1

Write two ways to break down the number. Possible answers are given.

78. 40

$2 \times 5 \times 2 \times 2; 5 \times 4 \times 2$

79. 32

$8 \times 2 \times 2; 4 \times 4 \times 2$

80. 45

$3 \times 3 \times 5; 5 \times 9$

81. 54

$2 \times 3 \times 9; 2 \times 3 \times 3 \times 3$

Write the missing factor.

82. $63 = \square \times 3 \times 7$

3

83. $36 = 2 \times 2 \times 3 \times \square$

3

NS 4.2

Write *prime* or *composite* for each number.

84. 19   prime

85. 38   composite

86. 56   composite

87. 43   prime

Write each as a product of prime factors.

88. 36 $2 \times 2 \times 3 \times 3$

89. 55 5×11

90. 58 2×29

91. 60 $2 \times 2 \times 3 \times 5$

Benchmark Assessment

Name _____

ALGEBRA AND FUNCTIONS

AF 1.0

Matt enters data on the keyboard at a rate of 25 words per minute. If he enters data for 8 minutes, about how many words has he written?

1. What equation can you use to answer the question?

    A $25 = 8 \times n$
    B $25 \times 8 = n$
    C $n = 8 \times 5$
    D $8 = 25 \times n$

2. What solution answers the question?

    A 50 words
    B 125 words
    C 175 words
    D 200 words

Write an equation and solve.

3. Kara makes 60 cookies. She fills the baking sheet 5 times. If she fits an equal number on each time, how many cookies are on the sheet?

    Possible answer:
    $60 \div 5 = d$; $d = 12$, where d is the number of cookies

4. Lee buys 9 tokens at the arcade. Each token costs $0.25. What is the total he spends?

    Possible answer:
    $9 \times \$0.25 = t$; $t = \$2.25$, where t is the total he spends

Name all possible whole number values for n.

5. $\frac{1}{2} = \frac{n}{6}$

    $n = 3$

6. $\frac{3}{4} > \frac{n}{2}$

    $n = 0, 1$

Compare. Write >, <, or = for each ◯.

7. $\frac{1}{3}$ ◯> $\frac{1}{6}$

8. $\frac{3}{8}$ ◯< $\frac{7}{8}$

9. $\frac{2}{3}$ ◯= $\frac{4}{6}$

10. $\frac{1}{2}$ ◯< $\frac{3}{4}$

11. 0.45 ◯> 0.4

12. 0.9 ◯< 0.98

13. 0.07 ◯< 0.77

14. 0.6 ◯= 0.60

Benchmark Assessment

Name _____

> MATHEMATICAL REASONING

MR 1.0

Find a compatible number for the dividend so you can estimate the quotient.

1. 275 ÷ 3

   Possible answers: 270, 300

2. 550 ÷ 9

   Possible answers: 540, 630

3. Phil estimated 414 ÷ 6 by rounding the dividend to 400. Is 400 a good choice? Explain.

   No; 400 is not compatible with 6. A better choice is 420.

4. Use estimation to tell which quotient is greater.

   742 ÷ 9 or 229 ÷ 3

   742 ÷ 9

MR 1.1

Solve.

5. Three friends count out 85 mini cookies and divide them equally. How many cookies does each get?

   A 28; drop the remainder
   B 29; increase the quotient by 1
   C 1; use the remainder as the answer
   D 28; there is no remainder

6. Pudding cups are sold in four-packs. How many packs should you buy to serve 29 students one pudding each?

   A 7; drop the remainder
   B 8; increase the quotient by 1
   C 1; use the remainder as the answer
   D 7; there is no remainder

Write the next two numbers in the following sequences.

7. 11, 14, 18, 23, 29, __, __

   36, 44

8. 1, 2, 3, 5, 7, 11, __, __

   13, 17 (prime numbers)

9. Find the mystery number. The two numbers before it are 49 and 56. The two numbers after it are 70 and 77. The number is less than 100.

   63

10. What information is irrelevant in problem 9?

    The number is less than 100.

Benchmark Assessment

Name _____

MATHEMATICAL REASONING

MR 1.2

11. Which choice shows 43 × 26 broken into simpler parts?

    A (6 × 3) + (6 × 40) + (20 × 3) + (20 × 40)
    B (6 × 40) + (2 × 23)
    C (40 × 20) + (3 × 6)
    D (6 × 3) + (6 × 4) + (20 × 3) + (20 × 40)

12. What is the product of 43 × 26?

    A 246
    B 818
    C 902
    D 1,118

Break into simpler parts and solve.

13. 95 × 31

    <u>　1　</u> × <u>　5　</u> = <u>　5　</u>
    <u>　1　</u> × <u>　90　</u> = <u>　90　</u>
    <u>　30　</u> × <u>　5　</u> = <u>　150　</u>
    <u>　30　</u> × <u>　90　</u> = <u>　2,700　</u>
    <u>　5　</u> + <u>　90　</u> + <u>　150　</u> +
    <u>　2,700　</u> = <u>　2,945　</u>

14. 57 × 78

    <u>　8　</u> × <u>　7　</u> = <u>　56　</u>
    <u>　8　</u> × <u>　50　</u> = <u>　400　</u>
    <u>　70　</u> × <u>　7　</u> = <u>　490　</u>
    <u>　70　</u> × <u>　50　</u> = <u>　3,500　</u>
    <u>　56　</u> + <u>　400　</u> + <u>　490　</u> +
    <u>　3,500　</u> = <u>　4,446　</u>

MR 2.1

Use estimation to determine whether the answer is reasonable. Explain.

15. Sara calculates her expenses at the party store will be $36.48. She is buying balloons ($5.25), plates ($2.50), cups ($1.85), and a new game ($16.88).

    No; round each cost to the nearest dollar. $5 + $3 + $2 + $17 = $27; the estimate is $27.

16. Bart calculates the cost of his lunch at $4.93. He pulls out a five-dollar bill to pay. He is buying a burger ($1.99), fries ($1.65), and a shake ($2.29). Is this reasonable?

    No; round each cost to the nearest dollar. $2 + $2 + $2 = $6; the estimate is $6.

Benchmark Assessment

Name _____

MATHEMATICAL REASONING

MR 2.3

Make a number line to solve.

17. Marco ate $\frac{3}{4}$ of his pizza, and Rob ate $\frac{5}{8}$ of his pizza. Who ate more?

   Marco: $\frac{3}{4} > \frac{5}{8}$

18. Four recipes to make gingerbread call for different amounts of molasses: $\frac{1}{2}$ cup, $\frac{1}{3}$ cup, $\frac{2}{3}$ cup, $\frac{1}{4}$ cup. Order the amounts from least to greatest.

   $\frac{1}{4}, \frac{1}{3}, \frac{1}{2}, \frac{2}{3}$

Make a model to solve. Simplify fractions.

19. Greta cut the circular cheesecake into 12 slices. She gave away $\frac{1}{4}$ and $\frac{1}{2}$ to her friends. What fraction was left? How many slices is this?

   $\frac{1}{4}$; 3 slices

20. Kurt needed $\frac{3}{4}$ yd of PVC pipe for one project and another $\frac{3}{4}$ yd for a different project. How much did he need altogether?

   $1\frac{1}{2}$ yd

21. Lynn swam for $\frac{5}{6}$ hour. Emma swam for $\frac{1}{6}$ hour. How much longer did Lynn swim than Emma?

   $\frac{4}{6}$, or $\frac{2}{3}$ hour

22. One football player throws the ball $\frac{7}{10}$ of the field. Another throws it $\frac{1}{2}$ of the field. How much farther does one player throw the ball than the other?

   $\frac{2}{10}$, or $\frac{1}{5}$ of the field

Make a model to solve.
A spinner has 6 equal sections. Two sections are red, 1 section is blue, and 3 sections are yellow.

23. Which correctly describes the spinner?

   A $\frac{2}{6}$ red, $\frac{2}{6}$ blue, $\frac{3}{6}$ yellow
   B $\frac{2}{6}$ red, $\frac{1}{6}$ blue, $\frac{1}{3}$ yellow
   C $\frac{1}{3}$ red, $\frac{1}{6}$ blue, $\frac{1}{2}$ yellow
   D $\frac{1}{4}$ red, $\frac{1}{6}$ blue, $\frac{1}{2}$ yellow

24. If the blue section is changed to yellow, what fraction of the spinner will be yellow?

   A $\frac{2}{3}$
   B $\frac{1}{2}$
   C $\frac{1}{3}$
   D $\frac{3}{4}$

Name _____

MATHEMATICAL REASONING

MR 2.6

Make a table or a model to organize data. Solve.

25. Jen, Ann, Lara, and Katie are standing in line. Katie is behind both Ann and Jen. Jen is behind Lara but ahead of Ann. Who is first in line?

    A Jen
    B Ann
    (C) Lara
    D Katie

26. A brown rabbit, white rabbit, and gray rabbit live in a forest, a pet store, and a backyard hutch, but not in that order. The white rabbit will be sold soon. The rabbit that finds his own food is not brown. Where does the brown rabbit live?

    (A) backyard hutch
    B pet store
    C forest
    D Not Given

Mike got results for 5 daily quizzes. He got an 89, 79, 81, 87, and 85, not in order. Monday was his highest score, and Thursday was the lowest. The score for Wednesday was lower than for Friday but higher than for Tuesday.

27. On which day did he get a score of 87?

    A Monday
    B Tuesday
    C Thursday
    (D) Friday

28. On which day was the median score for the daily quizzes?

    A Monday
    (B) Wednesday
    C Thursday
    D Friday

29. Ed, Chris, Alex, and John compare their heights. Ed is taller than John. John is taller than Chris but shorter than Alex. Ed is not the tallest. Who is tallest?

    Alex

30. Four game icons appear vertically on the computer desktop. The icon for the space game is directly under the one for the car game. The icon for the football game is above the one for the fishing game and below the one for the space game. Which game icon is at the bottom?

    fishing game icon

© Harcourt

Benchmark Assessment

Name _____

MATHEMATICAL REASONING

MR 2.6

31. Find the missing numbers in the pattern: 3, 32, 61, 90, ☐, ☐, 177

    _____119, 148_____

32. What are the next three colors in the sequence?

    red, red, green, blue, green, red, red, green, blue, green, red

    _____red, green, blue_____

33. At the first practice the team runs for 10 minutes. At each of the following practices, they add another 3 minutes. How long do they run at the seventh practice?

    _____28 minutes_____

34. What are the next two numbers in the sequence?

    33, 4, 34, 8, 35, 12, ☐, ☐

    _____36, 16_____

35. Kevin finished $\frac{3}{4}$ of his model car. Kyle completed $\frac{3}{8}$ of his model car. Who completed more?

    _____Kevin; $\frac{3}{4} > \frac{3}{8}$_____

36. Tavia spent $\frac{4}{6}$ of her money. Michelle spent $\frac{1}{3}$ of her money on clothes and $\frac{1}{3}$ on gifts. Which girl spent a greater fraction of her money?

    _____They spent the same amount; $\frac{4}{6} = \frac{2}{3}$._____

MR 3.1

37. Jason needed to buy display pages for 125 buffalo nickels. Each page could display 8 coins. How many pages should he buy? Explain.

    _____16; only 120 coins would fit on 15 pages_____

38. Nina made 47 cards to sell. She sold them in packages of six. How many packages could she sell?

    _____7, with 5 cards left over_____

39. Tyson bought a soccer ball for $16.49, soccer shoes for $39.29, and shin guards for $12.50. Is it reasonable to say that Tyson spent more than $50? Explain.

    _____Yes; $70 > $50_____

40. Shawna is buying strawberries for $2.50, bananas for $2.08, a pineapple for $2.49, and sherbet for $3.75. She says $10 will cover the cost. Is this reasonable? Explain.

    _____No; $11 > $10_____

Benchmark Assessment

Name _____

UNITS 4-6 • PAGE 1

Write the correct answer.

1. **NS 1.2** Order the amounts of rainfall from least to greatest.

Daily Rainfall (inches)	
Carolton	1.09
Sunnymount	1.10
Dunsdale	0.90
Chesterville	0.99

0.90; 0.99; 1.09; 1.10

2. **NS 1.4** Ms. Johnson would like to order one notebook for each of her students. There are 27 students in each of her 5 classes. What is a reasonable estimate for the number of notebooks she should order?

about 150

3. **NS 1.5** What part is shaded? What part is unshaded?

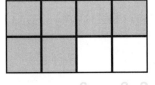

$\frac{6}{8}$, or $\frac{3}{4}$; $\frac{2}{8}$, or $\frac{1}{4}$

4. **NS 1.6** What is the decimal for $2\frac{7}{100}$?

2.07

Choose the letter of the correct answer.

5. **NS 1.2** Which group is ordered from **least** to **greatest**?

A 7.4, 7.09, 6.8, 6.65
B 7.09, 7.4, 6.65, 6.8
C 6.8, 6.65, 7.09, 7.4
D 6.65, 6.8, 7.09, 7.4

6. **NS 1.4** Karen gave $4.56 to her brother and $5.15 to each of her two sisters. Which is a good estimate for the amount of money she gave away?

A about $15.00
B about $10.00
C about $9.80
D about $0.15

Benchmark Assessment

Name _____

UNITS 4–6 • PAGE 2

Write the correct answer.

7. **NS 1.7** Write the word name for the model.

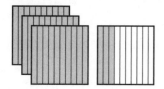

three and three tenths

8. **NS 1.9** Write a true statement comparing the two points on the number line. Use <, >, or =.

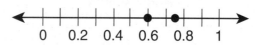

_0.75 > 0.60 or 0.60 < 0.75_

Choose the letter of the correct answer.

9. **NS 1.5** Which fraction names the shaded part?

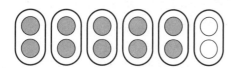

A $\frac{12}{12}$ C $\frac{4}{6}$
B $\frac{5}{6}$ D $\frac{5}{12}$

10. **NS 1.6** Which statement is **NOT** true?

A $0.25 = \frac{1}{4}$
B $\frac{1}{2} = 0.5$
C $0.75 = \frac{3}{4}$
D $0.02 = \frac{2}{10}$

11. **NS 1.7** Which fraction names the shaded part?

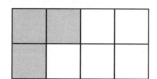

A $\frac{3}{8}$ C $\frac{6}{8}$
B $\frac{5}{8}$ D $\frac{7}{8}$

12. **NS 1.9** Which number line shows a fraction equivalent to $\frac{8}{12}$?

A

B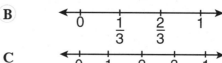

C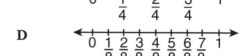

D

13. **NS 1.5** **Write About It** Explain how you solved Problem 9.

Student answers should include the following concepts: The shaded parts represent parts of a set. The set has 6 parts. Five parts are shaded. The fraction is $\frac{5}{6}$.

Name _____

UNITS 4–6 • PAGE 3

Write the correct answer.

14. **NS 2.0** What is 2.4 − 1.6?

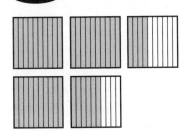

_____ 0.8 _____

15. **NS 2.1** One pen costs $1.35. Another pen costs $1.92. What is the total cost?

_____ $3.27 _____

Choose the letter of the correct answer.

16. **NS 2.0** 1.5 + 1.8

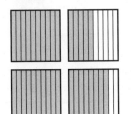

A 2.13
B 2.3
C 3.3
D Not Given

17. **NS 2.1** 2.65 − 0.93

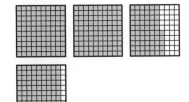

A 0.93
B 1.72
C 2.32
D 2.72

18. **NS 2.2** Which is 45.65 rounded to the nearest tenth?

A 50.00
B 40.00
C 45.00
D 45.70

19. **NS 3.0**   45,342
              − 37,999

A 6,343
B 7,343
C 13,443
D 17,343

Benchmark Assessment 45

Name _____

UNITS 4–6 • PAGE 4

Write the correct answer.

20. **NS 2.2** The sweater cost $23.69. What is the price rounded to the nearest dollar?

   $24

21. **NS 3.0** 2 × 40,000

   80,000

22. **NS 3.2** 7)6,124

   874 r6

23. **NS 3.3** A restaurant supply company bought 2,045 bottles of hot sauce for $0.79 a bottle. How much did they pay for the hot sauce?

   $1,615.55

Choose the letter of the correct answer.

24. **NS 3.2** Fifteen students are in a photography class. Each student took 1 roll of 36 photos and 2 rolls of 24 photos. Which expression will help you find how many photos were taken in all?

   A 15 × 1 × 36 × 2 × 24
   B (1 × 36) + (2 × 24) × 15
   C (2 × 24 × 36) × 15
   D (2 + 24 + 36) × 15

25. **NS 3.3** Fifty new computers were purchased for the school at a price of $895.90 each. What was the total cost?

   A $44,795.00
   B $45,795.00
   C $7,005.55
   D $447.95

46 Benchmark Assessment

Name _____

UNITS 4–6 • PAGE 5

Write the correct answer.

26. **NS 3.4** Howard bought his car 5 years ago. He has driven 62,340 miles. What is the average number of miles he drove each year?

_____12,468_____

27. **NS 4.0** What is the number 36 written as a product of prime factors?

_____2 × 2 × 3 × 3_____

Choose the letter of the correct answer.

28. **NS 3.4** Gary is planning a long bike trip. He knows he bikes about 9 miles in one hour. How many hours of biking will it take him to travel 504 miles?

A 7 hours
B 8 hours
C 56 hours
D 63 hours

29. **NS 4.0** Which is an equation for the arrays shown?

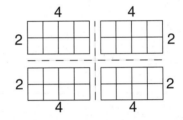

A 2 × (2 × 4) = 16
B 8 × 2 = 16
C 8 × (3 × 4) = 48
D 4 × (2 × 4) = 32

30. **NS 4.1** What are all the factors of 60?

A 2, 3, 4, 5, 6, 10
B 60, 120, 240, 480
C 1, 2, 3, 4, 6, 10, 20, 30, 60
D 1, 2, 3, 4, 5, 6, 10, 12, 15, 20, 30, 60

31. **NS 4.2** Which list contains only prime numbers?

A 3, 7, 11, 17
B 2, 3, 12, 19
C 2, 3, 5, 7, 15
D Not Given

Benchmark Assessment

Name _____

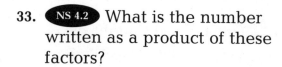

Write the correct answer.

32. **NS 4.1** What are all the factors of 56?

    1, 2, 4, 7, 8, 14, 28, 56

33. **NS 4.2** What is the number written as a product of these factors?

    $3 \times 3 \times 4 \times 2$

    72

34. **AF 1.0** Write <, >, or = to make a true statement.

    $\frac{3}{4} \bigcirc \frac{2}{3}$

    >

35. **MR 1.0** Terri has 48 tennis balls. She will give 4 balls to each of the students in her tennis class. How many students does Terri have in her class?

    12

Choose the letter of the correct answer.

36. **AF 1.0** Ron read the first 29 pages of a 245-page book. Which equation can be used to find the number of pages Ron still needs to read?

    A $245 + n = 29$
    B $n - 29 = 245$
    C $245 - 29 = n$
    D $245 + 29 = n$

37. **MR 1.0** Five students would like to divide 187 beads equally. Which is a reasonable estimate of how many beads each student will get?

    A between 10 and 20
    B between 20 and 30
    C between 30 and 40
    D between 40 and 50

Name _____

UNITS 4–6 • PAGE 7

Write the correct answer.

38. **MR 1.1** What are the missing numbers in the pattern?

    4, 8, 13, 19, ☐, ☐, 43

    _____26, 34_____

39. **MR 1.2** A theater has 30 rows of seats. Each row contains 18 seats. How many seats does the theater have?

    _____540_____

40. **MR 2.1** During the week Partap keeps track of the number of miles he drives. What is a reasonable estimate of the number of miles he drove in all?

MILES DRIVEN LAST WEEK	
Monday	9.9 mi
Tuesday	29.9 mi
Wednesday	58.9 mi
Thursday	20.2 mi
Friday	104.5 mi

    Possible answer: about 225 miles

41. **MR 2.3** Juan has a rectangular garden. He would like to plant $\frac{1}{4}$ of the garden with squash and $\frac{2}{8}$ of it with corn. How much of the garden does he have left to plant flowers?

    _____$\frac{1}{2}$_____

Choose the letter of the correct answer.

42. **MR 1.1** The juice box company donated 7 drinks for every 5 car wash volunteers. The car wash manager thinks there will be 75 volunteers. How many juice boxes will be donated?

    A 75 juice boxes
    B 105 juice boxes
    C 120 juice boxes
    D 525 juice boxes

43. **MR 1.2** Marnie is building a patio from 8-inch by 8-inch square stones. How many stones does she need to build a 96-inch by 96-inch patio?

    A 64
    B 96
    C 144
    D 150

44. **MR 1.1** **Write About It** Explain how you solved Problem 43.

    Student answers should include the following concepts: You can solve a problem by breaking it into simpler parts. Find how many 8-inch stones fit in 96 inches (96 ÷ 8 = 12 stones). Use the formula: $l \times w = a$ to find the total number of stones (12 × 12 = 144 stones).

Benchmark Assessment

Name _____

UNITS 4–6 • PAGE 8

Write the correct answer.

45. **MR 2.6** Ted has a lot of small wooden boards of the following sizes:

$\frac{2}{12}$ foot $\frac{3}{12}$ foot $\frac{5}{12}$ foot

Ted used three boards to make a 1-foot long section of board. What combination of boards did he use?

$\frac{5}{12} + \frac{5}{12} + \frac{2}{12}$

46. **MR 3.1** Kathy is mixing the following pony feed ingredients:

$\frac{1}{4}$ bucket oats

$\frac{1}{3}$ bucket bulgur

$\frac{1}{2}$ bucket crushed corn

How much feed does this make?

$1\frac{1}{12}$ buckets

Choose the letter of the correct answer.

47. **MR 2.1** Tate bought birdseed for $2.50, a birdcage for $21.95, three toys for her bird for $4 each, and a water bottle for $1.89. Which is a reasonable estimate of the total amount she spent?

  A $30
  B $35
  C $40
  D $50

48. **MR 2.3** Look at the figures below. The first figure is 2 × 2 and contains 4 squares. The second figure is 3 × 3 and contains 9 squares. The third figure is 4 × 4 squares and contains 16 squares. How many squares will Figure 10 contain?

Figure 1   Figure 2   Figure 3 . . . Figure 10

?

  A 100   B 121   C 144   D 169

49. **MR 2.6** Greg has 12 coins worth $1.00. The coins are either nickels or dimes. What combination of coins does Greg have?

  A 50¢ in dimes and 50¢ in nickels
  B 60¢ in dimes and 40¢ in nickels
  C 70¢ in dimes and 30¢ in nickels
  D 80¢ in dimes and 20¢ in nickels

50. **MR 3.1** Jake has the following items in his shopping cart: bread for $2.49, nut butter for $3.45, and jam for $2.59. Which of the following is the most reasonable estimate of his total?

  A $5.00
  B $6.00
  C $8.00
  D $10.00

Benchmark Assessment

Name _____

NUMBER SENSE

NS 1.8

Use the thermometer to find the temperature.

1. −17°F

2. −42°F

3. −24°C

For questions 4–9, use a thermometer to find the change in temperature.

4. 20°F to −10°F   30°

5. 5°F to −12°F   17°

6. −10°F to −17°F   7°

7. 40°C to −5°C   45°

8. −3°C to −15°C   12°

9. 37°C to −3°C   40°

For questions 10–17, use the number line.

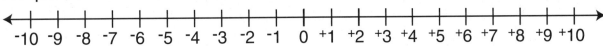

10. Order the integers from least to greatest: 0, −5, −10, −8

    −10, −8, −5, 0

11. Order the integers from least to greatest: 0, −2, 5, −7, 10

    −7, −2, 0, 5, 10

Compare. Write <, >, or = in each ◯.

12. ⁺5 > ⁻8

13. ⁻7 < ⁻4

14. ⁻3 < ⁺3

15. 0 > ⁻1

16. ⁺2 > ⁻4

17. ⁺6 < ⁺9

Benchmark Assessment   51

Name _____

ALGEBRA AND FUNCTIONS

AF 1.0

Change the units. Tell whether you multiply or divide.

1. 3 mi = ☐ ft   2. 288 in. = ☐ yd   3. 6 qt = ☐ c

   15,840; multiply   _8; divide_   _24; multiply_

4. ☐ oz = 5 lb   5. ☐ m = 400 cm   6. ☐ km = 6,000 m

   80; multiply   _4; divide_   _6; divide_

AF 1.1

Compare. Write <, >, or = for each ◯.

7. 45 ft ⊙(=) 15 yd   8. 2 T ⊙(<) 6,000 lb   9. 4 gal ⊙(>) 12 qt

10. 7 m ⊙(<) 7 km   11. 6 L ⊙(=) 6,000 mL   12. 3 kg ⊙(>) 2,000 g

For questions 13–14, choose the correct unit.

13. 2,000 g = 2 ___?___   14. 500 cm = 5 ___?___

    A m   **A m**
    B mL   B km
    C km   C dm
    D kg   D km

Solve.

15. Esteban takes 2 one-liter water bottles to his baseball game. At the end, he has half a bottle left. How many mL of water did he drink?

    _1,500 mL_

16. One serving of oat cereal is 30 g. How many servings does it take to make 3 kg?

    100 servings

Benchmark Assessment

Name _____

ALGEBRA AND FUNCTIONS

AF 1.4

Using the perimeter given, find the unknown length, *n*.

17. P = 328 yd

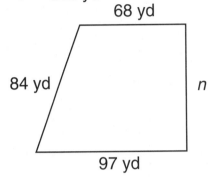

18. P = 72 cm

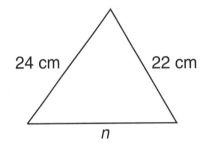

n = 79 yd

n = 26 cm

Write the formula for area or perimeter.
Find the unknown measurement.

19.

20.

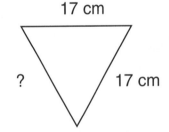

formula: $A = l \times w$; 20 m

formula: $P = a + b + c$; 17 cm

AF 1.5

For questions 21–22, use the equation $y = x - 3$.

21. Complete the function table.

Input	x	3	4	5	6	7
Output	y	0	1	2	3	4

22. Write the Input/Output values as ordered pairs (x,y).

(3,0), (4,1), (5,2), (6,3), (7,4)

Benchmark Assessment

Name _____

MEASUREMENT AND GEOMETRY

MG 1.0

Choose the correct answer.

1. An equilateral triangle has a perimeter of 42 cm. What is the length of one side?

   A 6 cm
   B 7 cm
   C 12 cm
   (D) 14 cm

2. Molly's greenhouse has an area of 54 square feet. The length is 9 feet. What is the width?

   (A) 6 ft
   B 18 ft
   C 19 ft
   D 36 ft

MG 1.1

Find the area.

3. Lee irons a tablecloth that measures 5 ft × 12 ft. What is its area?

   _____60 sq ft_____

4. Sara's beach towel measures 80 cm × 160 cm. What is its area?

   _____12,800 sq cm_____

MG 1.2 **MG 1.3**

For questions 5–8, choose the letter of the statement that is true.

   A Same perimeter and different area
   B Different areas and perimeters
   C Same area and perimeter
   D Same area but different perimeters

5. Rectangle X = 9 in. × 4 in.
   Rectangle Y = 18 in. × 2 in.

   _____D_____

6. Rectangle S = 4 ft × 12 ft
   Rectangle T = 7 ft × 9 ft

   _____A_____

7. Rectangle Q = 3 yd × 8 yd
   Rectangle R = 12 yd × 2 yd

   _____D_____

8. Rectangle C = 10 cm × 3 cm
   Rectangle D = 5 cm × 7 cm

   _____B_____

Benchmark Assessment

Name _____

MEASUREMENT AND GEOMETRY

MG 1.4

Find the area and perimeter of each floor plan.

9.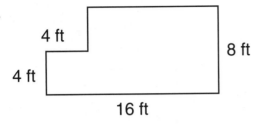

10.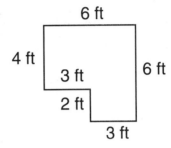

_A = 112 sq ft; P = 48 ft_

_A = 30 sq ft; P = 24 ft_

MG 2.0

For questions 11–13, use the coordinate grid.

11. Give the ordered pair for point C.

    (3,7)

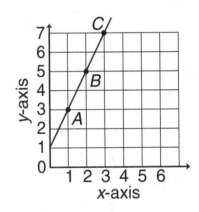

12. Give the ordered pair for point A.

    (1,3)

MG 2.1

13. The graph shows the equation $y = 2x + 1$.
    What is the value of y when $x = 2$?

    5

14. Which ordered pair would NOT make the equation $y = 2x + 1$ true?

    A (4,9)
    B (7,15)
    C (8,17)
    D (5,13)

15. Which ordered pair would NOT make the equation $y = 3x + 2$ true?

    A (0,2)
    B (1,6)
    C (2,8)
    D (3,11)

Benchmark Assessment

55

Name _____

MEASUREMENT AND GEOMETRY

MG 2.2

For questions 16–19, use the coordinate grid.

16. How many units long is the line segment joining the points (1,2) and (7,2)?

    _____6_____

17. Which two points have a horizontal distance of 3 units between them?

    A (1,2), (7,2)
    B (4,3), (4,5)
    C (1,5), (4,5)
    D (1,2), (1,5)

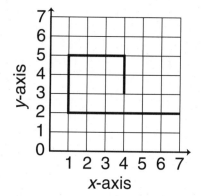

MG 2.3

18. How many units long is the line segment joining the points (1,2) and (1,5)?

    _____3_____

19. Which two points have a vertical distance of 2 units between them?

    _____(4,3), (4,5)_____

MG 3.0

For questions 20–21, use the figure.

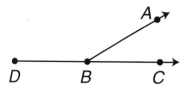

20. Which letter represents the vertex of angle *ABC*?

    _____B_____

21. Name the rays of angle *ABC*.

    _____BA, BC_____

56 Benchmark Assessment

Name _____

MEASUREMENT AND GEOMETRY

MG 3.1

For questions 22–23, use the map.

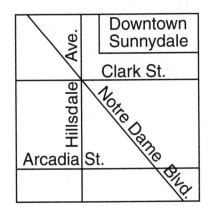

22. Which streets are parallel?

    Clark St., Arcadia St.

23. Which streets are perpendicular?

    Hillsdale Ave. and Clark St.,
    Hillsdale Ave. and Arcadia St.

NS 3.2

24. Name a radius and a diameter of the circle.

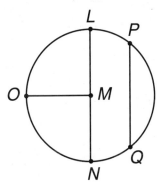

    $r =$ $\overline{MO}$, $\overline{ML}$, or $\overline{MN}$; $d =$ $\overline{LN}$

NS 3.3

25. Which shows congruent figures?

    A    B

    C   D

    B

MG 3.4

Lynn drew these pictures.

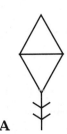      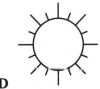
A   B   C   D

26. Which pictures show line symmetry?

    A, C, D

27. Which pictures show rotational symmetry?

    B, D

Benchmark Assessment   57

Name _____

MEASUREMENT AND GEOMETRY

MG 3.5

For questions 28–29, use the clock.

28. The clock shows 11:20. Which type of angle do the hands form: right, acute, or obtuse?

_____obtuse_____

29. By 12:05 the clock hands have moved $\frac{3}{4}$ of a complete turn. How many degrees of a circle does this represent?

_____270°_____

MG 3.6

30. Which of these figures has 5 vertices?

   A cube
   B triangular prism
   C triangular pyramid
   (D) square pyramid

31. Which figure can be made from this pattern?

   A cube
   B triangular prism
   (C) triangular pyramid
   D square pyramid

MG 3.7

For questions 32–33, classify each triangle by the length of its sides. Write *isosceles, scalene,* or *equilateral.*

32. 6 ft, 5 ft, 3 ft

_____scalene_____

33. 8 cm, 12 cm, 12 cm

_____isosceles_____

MG 3.8

34. Dan made a poster for student elections. It had one pair of parallel sides and no right angles. What was the figure he made?

_____trapezoid_____

35. Courtney used a table saw to make a cheese board. The board had two pairs of parallel sides, four congruent sides, and no right angles. What was the figure she made?

_____rhombus_____

STATISTICS, DATA ANALYSIS, AND PROBABILITY

Name _____

SDAP 1.0

For questions 1–3, use the tally table.

Number of Beads in the Can	
Outcome	Number of Times Pulled
blue	|||| |||| |||
yellow	|||| |||
green	|||| |||| ||||	
white	|||| |||| |||

1. Which color bead was pulled from the can most often?

   A blue
   B yellow
   C green
   D white

2. Which color bead probably has the fewest number in the can?

   yellow

3. Which color beads were pulled the same number of times?

   blue, white

For questions 4–5, make a chart. Solve.

4. Kayla packs for a trip. She includes three tops and two pairs of jeans. How many different outfits can she make?

   6

5. When one pair of jeans and one top are dirty, how many different clean outfits will Kayla have?

   2

For questions 6–7, make a list. Solve.

Pizza Galore offers choices for 5 toppings and 3 types of crusts.

Toppings	Crusts
mushrooms, olives, pepperoni, bacon, pineapple	thin, thick, stuffed

6. How many different pizzas that have only one topping can you order?

   15

7. If they add artichokes and sausage to the toppings list, how many different pizzas can you order that have only one topping?

   21

Benchmark Assessment

59

Name _____

STATISTICS, DATA ANALYSIS, AND PROBABILITY

SDAP 2.0

For questions 8–10, use the picture. Solve.

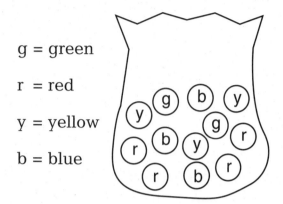

g = green
r = red
y = yellow
b = blue

8. Which color marble is least likely to be pulled from the bag?

   _____green_____

9. Which color marble is most likely to be pulled from the bag?

   _____red_____

10. Which color marbles are equally likely to be pulled from the bag?

    _____yellow and blue_____

A laundry basket holds for sorting: 10 white socks, 4 black socks, and 3 brown socks.

11. Without looking, which sock are you most likely to pick?

    (A) white
    B black
    C brown
    D green

12. Which sock is impossible to pick?

    A white
    B black
    C brown
    (D) green

SDAP 2.1

For questions 13–14, make a tree diagram.
Brent is using one spinner marked 5, 10, 15, 20 and a second spinner marked green, red, yellow, blue. Each is divided into equal fourths. Check student's diagrams.

13. How many outcomes are possible if you spin both spinners?

    A 4
    B 8
    C 12
    (D) 16

14. How many possible outcomes include green?

    A 1
    B 2
    (C) 4
    D 6

Benchmark Assessment

Name _____

STATISTICS, DATA ANALYSIS, AND PROBABILITY

SDAP 2.2

For questions 15–18, look at the spinner. Find the probability of each event.

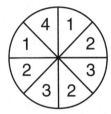

15. spin the number 2
    - A $\frac{2}{8}$, or $\frac{1}{4}$
    - **B** $\frac{3}{8}$
    - C $\frac{4}{8}$, or $\frac{1}{2}$
    - D $\frac{1}{8}$

16. spin the number 1
    - A $\frac{0}{8}$, or 0
    - B $\frac{1}{8}$
    - **C** $\frac{1}{4}$
    - D $\frac{3}{8}$

17. spin an even number

    $\frac{4}{8}$, or $\frac{1}{2}$
    _____

18. spin a prime number

    $\frac{5}{8}$
    _____

19. Joe tosses a number cube that is marked 1–6. What is the probability he will toss the number 5 on his second try?
    - A $\frac{6}{6}$, or certain
    - **B** $\frac{1}{6}$
    - C $\frac{2}{6}$, or $\frac{1}{3}$
    - D impossible

20. Drew and his friend take turns pulling a card from a bag. Drew will score 1 point if he pulls out a vowel, and his friend will score 1 point if he pulls out a consonant. Which letter cards will make a fair game?
    - A the letters for FAIRNESS
    - B the letters for GREAT TIME
    - C the letters for GOOD PALS
    - **D** the letters for FAIR GAME

21. Colleen has a bag with prizes to hand out. She has 12 whistles, 6 lanyards, and 12 kaleidoscopes. What is the probability, without looking, that she will pick a lanyard from the bag?

    $\frac{1}{5}$
    _____

22. Chung picks red, and Ana picks blue. The spinner has 3 red sections and 2 blue sections. What is the probability that Chung will score a point? that Ana will score a point? Will the game be fair?

    $\frac{3}{5}$, $\frac{2}{5}$; no
    _____

Benchmark Assessment

Name _____

> **MATHEMATICAL REASONING**

MR 1.0

Choose the most reasonable unit of measure.
Write *cm*, *dm*, *m*, or *km*.

1. the length of a moth

   _____cm_____

2. the length of a hockey rink

   _____m_____

MR 1.1

Decide if the problem has too much or too little information.
Solve, if possible.

3. A bus travels 15 kilometers in 10 minutes. How fast is it driving?

   ____too little information____

MR 2.3

4. Describe the different ways you can build a rectangular prism with a volume of 6 cubic units?

   ____6 × 1 × 1; 3 × 2 × 1____

MR 2.5

5. Which of the following requires an exact measurement instead of an estimate?

   A string to attach to a balloon
   B ribbon to make a bow
   C towel to use at the beach
   D frame for a photograph

MR 2.6

6. Taylor's team is picking new shirts. They can pick a print or solid color, in orange, green, silver, or black. How many choices do they have?

   A 8
   B 6
   C 5
   D 4

MR 3.2

7. What combinations of containers will make a quart?

   ____Possible answers: 1 pint and 2 cups; 4 cups____

MR 3.3

8. On Monday the heat index was 97°F, because the temperature was 85°F and the relative humidity was 80%. On Tuesday the temperature was the same, but the heat index was 102°F. Explain the change.

   ____The humidity was higher.____

Benchmark Assessment

Name _____

UNITS 7–9 • PAGE 1

Write the correct answer.

1. **NS 1.8** At 8:00 P.M. the temperature was 0°C. By midnight it had dropped 4°C. What was the temperature at midnight?

   _____−4°C_____

2. **AF 1.0** Bruce's sandbox measures $2\frac{2}{3}$ yards long. How many inches is this?

   _____96 inches_____

3. **NS 1.8** At 8:00 P.M., the temperature was ⁻1°C. By midnight, it had dropped to ⁻5°C. By how many degrees did the temperature change between 8:00 P.M. and midnight?

   _____4°C_____

4. **AF 1.0** Jessica is planning a party. There will be 16 people at the party. How many quarts of punch should she make so that everyone will be able to have 1 cup of punch?

   | 1 quart | = | 2 pints |
   | 1 pint | = | 2 cups |

   _____4 qt_____

Choose the letter of the correct answer.

5. **NS 1.8** Which number sentence is true?

   A ⁻5 > ⁻4
   B 3 < ⁻9
   C ⁻25 > 2
   (D) ⁻16 < ⁻7

6. **AF 1.0** 2L = ■ mL

   (A) 2,000
   B 200
   C 20
   D 10

7. **MG 1.4** Sarah wants to make a picture frame. She needs a 3-foot strip of wood for each side of the frame. The wood strip costs $2 a foot. How much money does she need to buy the wood for the frame?

   A $6   B $12   C $18   (D) $24

Benchmark Assessment

Name _____

UNITS 7–9 • PAGE 2

Write the correct answer.

8. **AF 1.1** The table shows the temperature in 4 cities. In which city is it coldest?

Chicago	Denver	Miami	New York
⁻16°F	⁻6°F	81°F	42°F

_____Chicago_____

9. **AF 1.4** Mr. Gomez put a fence around a square garden. He used 36 feet of fencing. What is the length of each side of his garden?

_____9 ft_____

10. **AF 1.5** What value completes the function table?
$$y = 2x - 1$$

Input, x	1	3	5
Output, y	1	5	

_____9_____

11. **AF 1.4** The perimeter of a card is 22 inches. If the width of the card is 5 inches, what is the length?

_____6 inches_____

Choose the letter of the correct answer.

12. **MG 2.0** Which ordered pair shows the location of the baseball field?

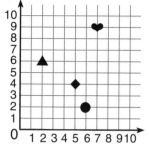

A (2,6)
B (4,5)
C (5,4)
D (6,2)

13. **MR 1.0** Write About It Explain how you solved Problem 11.

Student answers should include the following concepts:

Perimeter = 2l + 2w. P = 22 inches, so 22 = 2l + 10. Solve for l.

22 = 2l + 10; 22 − 10 = 2l + 10 − 10; 12 = 2l; 6 = l

Name _____

UNITS 7–9 • PAGE 3

Write the correct answer.

14. **MG 1.0** What is the perimeter of this rectangle?

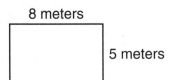

26 meters

15. **MG 2.2** What is the length of each side of the square?

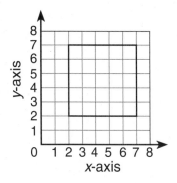

5 units

16. **MG 1.1** The dimensions of Stuart's garden are 5 feet by 3 feet. What is the area of his garden?

15 sq ft

17. **MG 1.2** Stuart wants to put a fence around his garden. It is 5 feet long and 3 feet wide. How much fencing will he need?

16 ft

Choose the letter of the correct answer.

18. **MG 2.1** Which ordered pair would **NOT** make the equation true?

$y = 2x + 4$

A (0,4)
B (1,6)
C (2,7)
D (3,10)

19. **MG 1.3** Which is true about these figures? They have—

10 cm, 4 cm 8 cm, 6 cm

A same perimeters, different areas.
B same areas, same perimeters.
C different areas, different perimeters.
D same areas, different perimeters.

Benchmark Assessment

Name _____

UNITS 7–9 • PAGE 4

Write the correct answer.

20. **MG 2.3** How many units long is the line segment joining the points (2,1) and (2,4)?

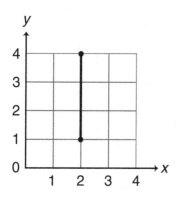

_____3 units_____

21. **MG 3.0** What does this solid figure best represent?

_____a cube_____

Choose the letter of the correct answer.

22. **MG 3.0** How many faces does the solid figure have?

A 7
B 6
C 5
D 4

23. **MG 3.1** Which street is NOT perpendicular to Street 2?

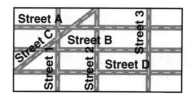

A Street A
B Street B
C Street C
D Street D

24. **MG 3.2** Which line segment is a diameter of the circle?

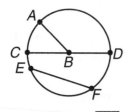

A $\overline{AB}$
B $\overline{BC}$
C $\overline{BD}$
D $\overline{CD}$

25. **MG 3.3** Which figures are congruent?

A

B

C

D Not Given

Benchmark Assessment

Name _____

UNITS 7-9 • PAGE 5

Write the correct answer.

26. **MG 3.6** Which net will form a cube?

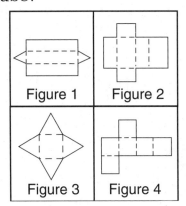

Figure 4

27. **MG 3.5** What type of angle is formed by the intersection of these lines?

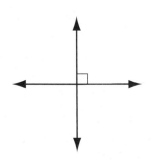

right angle

Choose the letter of the correct answer.

28. **MG 3.7** Classify the triangle.

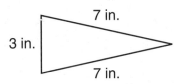

A equilateral   C isosceles
B scalene   D Not Given

29. **MG 3.5** Which word best describes angle B?

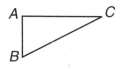

A right   C obtuse
B acute   D scalene

30. **MG 3.4** Which does **NOT** show a figure with a line of symmetry?

A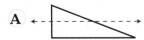

B

C

D

31. **MG 3.8** Which of these figures is **NOT** a parallelogram?

A

B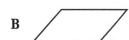

C

D

Benchmark Assessment

Name _____

UNITS 7–9 • PAGE 6

Write the correct answer.
Use the tally table for
Problems 32 and 33.

Number of Blocks in a Bag	
Outcome	Number of Times Pulled
White	𝍷𝍷𝍷𝍷𝍷𝍷𝍷𝍷𝍷𝍷𝍷𝍷𝍷𝍷𝍷𝍷𝍷
Black	|||
Red	𝍷𝍷𝍷𝍷𝍷𝍷𝍷
Green	𝍷𝍷𝍷𝍷𝍷𝍷𝍷𝍷𝍷𝍷𝍷𝍷𝍷𝍷𝍷𝍷𝍷𝍷𝍷𝍷𝍷𝍷

32. SDAP 1.1 Which color block was pulled from the bag most often?

_____green_____

33. SDAP 2.0 Which color block do you predict has the fewest number in the bag?

_____black_____

34. SDAP 2.0 A bag holds 12 marbles. Of them, 3 are clear, 4 are black, and 5 are purple. Without looking, which color marble are you most likely to pick?

_____purple_____

35. SDAP 2.2 Jessie has a bag of jelly beans. There are 15 red, 10 blue, 8 purple, and 7 white. If she picks a bean, without looking, what is the probability that she will pick a purple?

_____$\frac{1}{5}$_____

Choose the letter of the correct answer.

36. SDAP 2.1 Alfredo bought these clothes. How many different pants and shirt combinations could he make?

A 2
B 6
C 8
D 42

37. SDAP 2.0 There are 6 balls in a bag. One ball is red, 2 are blue, and 3 are green. If Lisa pulls 1 ball without looking, what is the probability that it will be green?

A $\frac{1}{6}$ C $\frac{1}{3}$

B $\frac{1}{4}$ D $\frac{1}{2}$

68   Benchmark Assessment

Name _____

UNITS 7–9 • PAGE 7

Write the correct answer.

38. **MR 1.1** Mai is packing a box of gifts. She spent $10.95 on each of her 4 cousins. The box she is packing measures 8 inches high, 10 inches wide, and 12 inches long. What is the volume of the box?

960 cubic inches

39. **MR 2.3** Sam built a fence around a square vegetable garden. He put up 32 feet of fencing. What is the length of each side of his vegetable garden?

_8 ft_

Choose the letter of the correct answer.

40. **MR 2.3** Which figure is a quadrilateral?

A

B

C

D Not Given

41. **MR 1.0** Which is the most reasonable estimate of the mass of the nail?

A) 2 grams
B 2 kilograms
C 2 tons
D 2 meters

42. **SDAP 2.1** Every day John eats a sandwich, a banana, and yogurt for lunch. How many different ways could he arrange the order in which he eats the 3 items?

A 3
B 4
C) 6
D 8

43. **SDAP 2.2** A box has 4 red balls, 3 orange balls, and 1 blue ball. Which is **NOT** a possible outcome if 2 balls are drawn at the same time?

A drawing 2 reds
B drawing 2 oranges
C drawing a red and a blue
D) drawing 2 blues

Benchmark Assessment

Write the correct answer.

44. **MR 2.3** Which point names the ordered pair (4,1)?

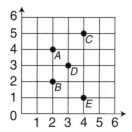

Point E

45. **MR 2.3** Is Saturday's temperature warmer or colder than Friday's temperature? How much warmer or colder?

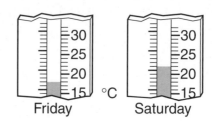

warmer; 4°

46. **MR 3.2** A book has a mass of 405 grams. About how much do 5 of these books weigh?

about 2,000 g, or 2 kg

47. **MR 3.2** The perimeter of rectangle WXYZ is 44 feet. What is the length of $\overline{XY}$?

10 ft

Choose the letter of the correct answer.

48. **MR 2.5** Find the average length of Ted's long jump, to the nearest inch. His jumps are 39 in., 47 in., and 42 in.

  A 39 in.   **C** 42 in.
  B 40 in.   **D** 43 in.

49. **MR 2.6** A cube is numbered 1, 3, 3, 5, 5, and 7. What is the probability of rolling 1?

  A 0   **C** $\frac{1}{3}$
  B $\frac{1}{6}$ **D** $\frac{1}{2}$

50. **MR 3.3** **Write About It** Use the weather information in the table. What generalizations can you make about the two cities?

WEATHER IN CALIFORNIA CITIES

City	Normal Temp. in Jan. (°F)	Normal Temp. in July (°F)	Precipitation in Jan. (in.)	Precipitation in July (in.)
San Diego	57	71	1.8	<0.05
San Francisco	49	63	4.4	<0.05

Possible answer: Temperatures are lower in San Francisco than in San Diego. San Francisco than has more rain than San Diego.

Name _____

PRACTICE TEST 1 • PAGE 1

Write the correct answer.

1. **NS 1.1** Write the number three million, five hundred twenty-six thousand, two hundred forty in standard form.

   _____3,526,240_____

2. **NS 1.2** Write the numbers in order from **least** to **greatest**:

   14,698; 13,567; 14,654

   _____13,567; 14,654; 14,698_____

Choose the letter of the correct answer.

3. **NS 1.3** A census counted the population of Kenya as 28,817,227 people. What is 28,817,227 rounded to the nearest hundred thousand?

   A 30,000,000
   B 29,000,000
   C 28,800,000
   D 28,000,000

4. **NS 1.4** Four children equally share $49. What is a reasonable estimate of how much money each receives?

   A about $14
   B about $12
   C about $10
   D about $9

5. **NS 1.5** Which mixed number does the picture show?

   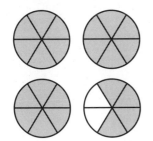

   A $3\frac{4}{24}$ C $3\frac{2}{3}$
   B $3\frac{2}{5}$ D $4\frac{4}{6}$

6. **NS 1.6** Which decimal is equivalent to the fraction $\frac{36}{100}$?

   A 36.00
   B 3.60
   C 3.06
   D 0.36

Benchmark Assessment

Name _____

PRACTICE TEST 1 • PAGE 2

Write the correct answer.

7. **NS 1.7** Marco makes 64 baskets out of 100 free throws. Write the fraction that represents Marco's free throws.

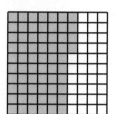

_____ $\frac{64}{100}$ _____

8. **NS 1.8** The weather person said that the temperature was "twenty-two degrees below zero." How do you write this number?

_____ $-22°$ _____

Choose the letter of the correct answer.

9. **NS 1.9** Which number is greater than $\frac{13}{3}$? Use the number line.

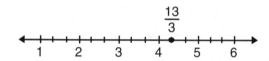

A $4\frac{2}{3}$ C $3\frac{1}{3}$

B $4\frac{1}{3}$ D $\frac{9}{3}$

10. **NS 2.1** Martha had $44.79. She spent $39.14 on a sweater. How much money did she have left?

A $83.93   C $15.65
B $80.00   D $5.65

11. **NS 2.2** The fourth-grade class had a bake sale. They made $24.09 selling cookies and $15.92 selling cakes. What is a reasonable estimate of the total sales?

A $9   C $40
B $30   D $50

12. **NS 3.1** 450,788 − 398,009

A 14,079
B 52,779
C 848,797
D Not Given

13. **NS 2.2** **Write About It** Explain how you chose a reasonable estimate for Problem 11.

Possible answer: I rounded $24.09 to $20.00 and $15.92 to $20.00.

$20.00 + $20.00 = $40.00, which makes $40.00 a reasonable estimate.

Benchmark Assessment

Name _____

PRACTICE TEST 1 • PAGE 3

Write the correct answer.

14. **NS 3.2** What value of *n* makes the equation true?

    $11 \times n = 77$

    _____ $n = 7$ _____

15. **NS 3.3** Mr. Gessner earns $1,289 a week. Last year he worked 49 weeks. How much money did Mr. Gessner earn?

    _____ $63,161 _____

Choose the letter of the correct answer.

16. **NS 3.4** The Hamilton School System has 1,015 students in 5 schools. If each school has the same number of students, how many students are in each school?

    A 23
    B 203
    C 2,003
    D 5,075

17. **NS 4.1** Which lists all the factors of 40?

    A 1, 2, 4, 5
    B 1, 4, 5, 8, 10, 40
    C 1, 10, 20, 40
    D 1, 2, 4, 5, 8, 10, 20, 40

18. **NS 4.2** Which set of numbers contains only prime numbers?

    A 5, 9, 15
    B 13, 17, 39
    C 5, 13, 21
    D 7, 11, 19

19. **AF 1.1** Lisa read 68 pages of a 215-page book. Which equation can be used to find the number of pages Lisa still needs to read?

    A $215 + p = 68$
    B $p - 68 = 215$
    C $215 - 68 = p$
    D $215 + 68 = p$

Benchmark Assessment

73

Name _____

PRACTICE TEST 1 • PAGE 4

Write the correct answer.

20. **AF 1.2** What is the value of the expression?

    $(4 \times 5) \div (5 - 4)$

    _____20_____

21. **AF 1.4** The media center has an area of 108 square yards. How wide is the media center if the length is 9 yards?

    _____12 yd_____

Choose the letter of the correct answer.

22. **AF 1.3** Which expression has a value of 7?

    A $15 - (14 - 6)$
    B $(15 - 4) - 5$
    C $(16 - 8) + 2$
    D Not Given

23. **AF 1.5** Which ordered pair makes this equation true?

    $y = (x \div 6) - 1$

    A (2,18)
    B (18,2)
    C (24,4)
    D Not Given

24. **AF 2.1** Which equation is true for $n = 6$?

    A $8 - 6 = n - 5$
    B $6 + 6 = n + 5$
    C $n + 6 = 4 + 8$
    D $12 - n = 9 - 2$

25. **AF 2.2** What is the value of n if $n \div 5 = 14$?

    A 2
    B 9
    C 19
    D 70

Benchmark Assessment

Name _____

PRACTICE TEST 1 • PAGE 5

Write the correct answer.

26. **MG 1.1** Find a rectangle that has the same area but a different perimeter than this rectangle.

   6 ft
   [rectangle] 4 ft

   Student answers may include:

   2 ft x 12 ft, 3 ft x 8 ft, 4 ft x 6 ft,

   8 ft x 3 ft, or 12 ft x 2 ft

27. **MG 1.2** A rectangle has an area of 30 square inches. What might its length and width be?

   Student answers may include:

   1 in. x 30 in., 2 in. x 15 in.,

   3 in. x 10 in., 5 in. by 6 in.,

   6 in. x 5 in., 10 in. x 3 in.,

   15 in. x 2 in., or 30 in. x 1 in.

Choose the letter of the correct answer.

28. **MG 1.3** A rectangle has a perimeter of 48 centimeters. What might its length and width be?

   A 3 cm by 16 cm
   B 4 cm by 12 cm
   C 8 cm by 6 cm
   D 8 cm by 16 cm

29. **MG 1.4** A rectangular apricot orchard has a perimeter of 18 miles. The width is 3 miles. What is the length?

   A) 6 mi
   B 10 mi
   C 15 mi
   D 21 mi

30. **MG 2.1** What ordered pair makes the equation true?

   $y = 2x + 7$

   A (20,3)
   B (13,3)
   C (3,13)
   D (3,20)

31. **MG 2.2** Which two points have a horizontal distance of 2 units between them?

   A (1,8) and (7,8)
   B (5,3) and (3,3)
   C (0,6) and (8,6)
   D (2,9) and (5,9)

Benchmark Assessment

75

Name _____

PRACTICE TEST 1 • PAGE 6

Write the correct answer.

32. **MG 3.2** Use the circle below. Name each radius.

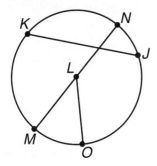

$\overline{LN}, \overline{LO}, \overline{LM}$

33. **MG 3.5** What type of angle is shown?

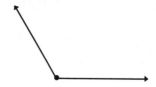

obtuse

Choose the letter of the correct answer.

34. **MG 2.3** Which two points have a vertical distance of 5 units between them?

    A (4,3) and (4,6)
    B (5,2) and (5,4)
    C (0,9) and (0,8)
    D Not Given

35. **MG 3.4** Which letter does **NOT** have line symmetry?

    A C
    B E
    C M
    D N

36. **MG 3.3** Which pair of figures is congruent?

    A
    B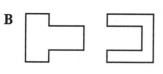
    C
    D Not Given

37. **MG 3.1** Which two lines are perpendicular?

    A
    B
    C
    D Not Given

Name _____

PRACTICE TEST 1 • PAGE 7

Write the correct answer.

38. **MG 3.6** When Tim folds this net, what solid figure does he make?

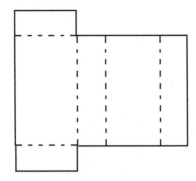

rectangular prism

39. **SDAP 1.1** Look at the graph. How many students chose Wild Berry?

Each 🍦 stands for 2 students.

4

Choose the letter of the correct answer.

40. **MG 3.7** This triangle is isosceles. Which could be the lengths of the sides?

A 4 ft, 5 ft, 5 ft
B 3 cm, 7 cm, 9 cm
C 6 ft, 6 ft, 6 ft
D 3 mm, 4 mm, 5 mm

41. **MG 3.8** Which terms best describe the figure below?

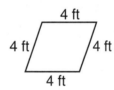

A square, quadrilateral
B rhombus, quadrilateral
C rectangle, quadrilateral
D trapezoid, quadrilateral

42. **SDAP 1.2** Marc sells tennis rackets at a sports store. On Monday he sold 13, on Tuesday he sold 14, and on Wednesday he sold 8. On the last two days of the week he sold 10 rackets each day. What is the mean number of rackets Marc sold?

A 6   C 11
B 10   D 12

43. **SDAP 1.3** The graph shows the number of goals scored by the soccer team. How many more goals were scored in the seventh game than in the first game?

A 1
B 2
C 3
D 4

Benchmark Assessment

Name _____

PRACTICE TEST 1 • PAGE 8

Write the correct answer.

44. **SDAP 2.1** The cube is numbered 1–6. How many possible outcomes are there if you toss the number cube and spin the pointer?

_____18_____

45. **MR 3.1** Jonah rode his bike for 1 hour 39 minutes. He returned home at 3:15 P.M. When did Jonah leave home?

_____1:36 P.M._____

Choose the letter of the correct answer.

46. **SDAP 2.2** What is the probability of rolling a 4 or 6 on a number cube labeled 1–6?

A $\frac{1}{6}$ C $\frac{4}{6}$
B $\frac{2}{6}$ D 1

47. **MR 2.2** Jason buys a sweatshirt for $39.05, shorts for $23.50, and a notebook for $14.95. Which is the most reasonable estimate of how much change he will receive from a $100 bill?

A $80   C $30
B $77.50 D $20

48. **MR 2.1** What is the area of the figure below?

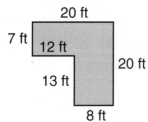

A 92 sq ft   C 244 sq ft
B 156 sq ft   D Not Given

49. **MR 1.2** For a crafts project, Alex needs to place 347 buttons on each of 8 tables. Which expression could she use to find the total number of buttons needed?

A (300 × 8) + (40 × 8) + (7 × 8)
B (300 × 8) × (40 × 8) × (7 × 8)
C 347 + 8
D (340 × 8) − (7 × 8)

50. **MR 2.4** **Write About It** Explain how you solved Problem 45.

Student answers may include either method: counting backward

from 3:15 P.M. or renaming hours as minutes and subtracting.

Students may also check their work by counting up or using addition.

Name _____

PRACTICE TEST 2 • PAGE 1

Write the correct answer.

1. **NS 1.1** Write the number sixty million, seven hundred fifty-two thousand, three hundred sixty in standard form.

   _____60,752,360_____

2. **NS 1.2** Write the numbers in order from **least** to **greatest**: 24,476; 23,654; 24,234

   _____23,654; 24,234; 24,476_____

Choose the letter of the correct answer.

3. **NS 1.3** A census counted the population of Egypt as 62,359,623 people. What is 62,359,623 rounded to the nearest thousand?

   A 63,000,000
   B 62,400,000
   C 62,369,000
   D 62,360,000

4. **NS 1.4** Four children equally share $27. What is a reasonable estimate of how much money each child receives?

   A about $17
   B about $10
   C about $7
   D about $5

5. **NS 1.5** Which number does the picture show?

   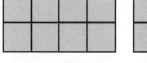
   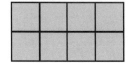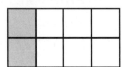

   A $24\frac{2}{8}$
   B $3\frac{6}{8}$
   C $3\frac{2}{8}$
   D $\frac{6}{8}$

6. **NS 1.6** Which decimal is equivalent to the fraction $\frac{15}{100}$?

   A 0.015
   B 0.15
   C 1.05
   D 1.5

Benchmark Assessment

Name _____

PRACTICE TEST 2 • PAGE 2

Write the correct answer.

7. **NS 1.7** Quality Control checks 3 out of every 100 teabags made. Write the fraction of teabags checked.

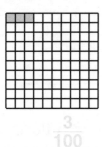

$\frac{3}{100}$

8. **NS 1.8** The radio announcer said that the temperature was "ten degrees below zero Fahrenheit." How is this number written?

⁻10°F

Choose the letter of the correct answer.

9. **NS 1.9** Which shows the fractions ordered from **least** to **greatest**?

0   $\frac{1}{2}$ $\frac{3}{4}\frac{7}{8}$ 1

A $\frac{3}{4} < \frac{1}{2} < \frac{7}{8}$ C $\frac{1}{2} < \frac{7}{8} < \frac{3}{4}$

B $\frac{3}{4} < \frac{7}{8} < \frac{1}{2}$ D $\frac{1}{2} < \frac{3}{4} < \frac{7}{8}$

10. **NS 2.1** Joel had $32.68. He spent $27.13 on a baseball. How much money did he have left?

A $5.55
B $15.55
C $59.81
D $69.81

11. **NS 2.2** What is a reasonable estimate for 62.87 + 17.04?

A 40   C 70
B 46   D 80

12. **NS 3.1** 700,034 − 592,349

A 106,685
B 107,625
C 107,685
D Not Given

13. **NS 2.2** **Write About It** Explain how you chose a reasonable estimate for Problem 11.

Student answers should include the following concepts:
Rounding two-place decimals to the nearest whole number, then rounding the whole number to the nearest ten; 62.87 rounds to 63. 17.04 rounds to 20. 60 + 20 = 80. So, 80 is a reasonable estimate.

Name _____

PRACTICE TEST 2 • PAGE 3

Write the correct answer.

14. **NS 3.2** What value of t makes the equation true?

$12 \times t = 60$

_____ $t = 5$ _____

15. **NS 3.3** The school pays its coaching staff $1,008 a week. Last year the coaching staff worked 32 weeks. How much money did the coaching staff earn?

_____ $32,256 _____

Choose the letter of the correct answer.

16. **NS 3.4** Greymount Publishing Company packed 6,309 test booklets in 9 cartons. If the same amount is in each carton, how many test booklets are in each carton?

A 7
B 71
C 701
D 901

17. **NS 4.1** Which shows all the factors of 36?

A 1, 4, 9, 36
B 1, 2, 3, 4, 6, 9, 18, 36
C 1, 6, 36, 216
D 1, 2, 3, 4, 5, 6

18. **NS 4.2** Which set of numbers contains only prime numbers?

A 11, 13, 19
B 15, 17, 23
C 5, 7, 12
D 13, 19, 33

19. **AF 1.1** Tara ran 15 miles a week to train for a race. She ran 450 miles in all. Which equation can be used to find the number of weeks, w, Tara ran to train?

A $450 \div 52 = w$
B $w + 15 = 450$
C $450 - 15 = w$
D $15 \times w = 450$

Benchmark Assessment

Name _____

PRACTICE TEST 2 • PAGE 4

Write the correct answer.

20. **AF 1.2** What is the value of this expression?

    $(89 - 26) \div (5 + 2)$

    _____9_____

21. **AF 1.4** The length of a rectangular field is 25 yards. Its perimeter is 80 yards. What is the width of the field?

    _____15 yd_____

Choose the letter of the correct answer.

22. **AF 1.3** Which expression has the same value as $(5 \times 5) - 4$?

    A $(7 \times 5) - 2$
    B $5 \times (5 - 4)$
    C $7 \times (5 - 2)$
    D Not Given

23. **AF 1.5** Which ordered pair makes this equation true?

    $(x \times 3) + 1 = y$

    A (3,10)
    B (10,3)
    C (4,8)
    D (8,4)

24. **AF 2.1** Which equation is true for $z = 2$?

    A $z + 7 = 3 + 6$
    B $7 + 3 = z + 2$
    C $7 - 5 = z - 1$
    D $12 + z = 13 + 2$

25. **AF 2.2** What is the value of n if $n \div 7 = 16$?

    A 2 r2
    B 9
    C 23
    D 112

82   **Benchmark Assessment**

Name _____

PRACTICE TEST 2 • PAGE 5

Write the correct answer.

26. **MG 1.1** How much more area does a 6-foot by 6-foot square cover than a 4-foot by 6-foot rectangle?

    _____12 square feet_____

27. **MG 1.2** A rectangle has an area of 16 square inches. What might its length and width be?

    Student answers may include:

    1 in. × 16 in., 2 in. × 8 in.,

    4 in. × 4 in., 8 in. × 2 in.,

    16 in. × 1 in.

Choose the letter of the correct answer.

28. **MG 1.3** A rectangle has a perimeter of 32 centimeters. What might its length and width be?

    A 2 cm by 16 cm
    B 8 cm by 4 cm
    C 8 cm by 8 cm
    D 8 cm by 24 cm

29. **MG 1.4** A rectangular pad has an area of 42 square centimeters. The width is 7 centimeters. What is its perimeter?

    A 6 cm
    B 13 cm
    C 19 cm
    D 26 cm

30. **MG 2.1** What ordered pair makes the equation true?

    $y = 3x - 4$

    A (4,16)
    B (4,3)
    C (3,4)
    D (4,8)

31. **MG 2.2** Which two points have a horizontal distance of 6 units between them?

    A (1,3) and (6,3)
    B (1,8) and (9,8)
    C (0,4) and (6,4)
    D (3,2) and (8,2)

Benchmark Assessment

Name _____

PRACTICE TEST 2 • PAGE 6

Write the correct answer.

32. **MG 3.2** The center of this circle is G. Name each radius.

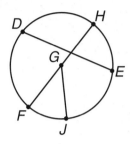

_____ GF, GJ, GH _____

33. **MG 3.5** What type of angle is shown?

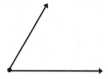

_____ acute _____

Choose the letter of the correct answer.

34. **MG 2.3** Which two points have a vertical distance of 4 units between them?

   A (4,1) and (4,6)
   B (0,7) and (0,1)
   C (5,2) and (5,3)
   D (7,2) and (7,6)   ⟵ circled

35. **MG 3.4** Which letter does NOT have line symmetry?

   A Z
   B T
   C B
   D X

36. **MG 3.3** Which pair of figures is congruent?

   A

   B      ⟵ circled

   C

   D Not Given

37. **MG 3.1** Which two lines appear to be parallel?

   A

   B

   C   ⟵ circled

   D Not Given

Name _____

PRACTICE TEST 2 • PAGE 7

Write the correct answer.

38. **MG 3.6** When Niku folds this net, what solid figure does he make?

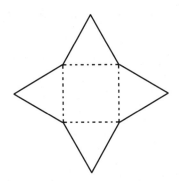

square pyramid

39. **SDAP 1.1** The graph shows the favorite types of fruit of a group of people. How many people chose apples as their favorite type of fruit?

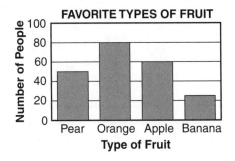

60

Choose the letter of the correct answer.

40. **MG 3.7** This triangle is scalene. Which could be the lengths of the sides?

A 3 ft, 3 ft, 9 ft
B 4 m, 5 m, 5 m
C 4 cm, 3 cm, 2 cm
D 3 yd, 3 yd, 3 yd

41. **MG 3.8** Which best describe the figure below?

A square, quadrilateral
B rhombus, quadrilateral
C rectangle, quadrilateral
D parallelogram, quadrilateral

42. **SDAP 1.2** Kelley School had these absences for one week:

Monday: 8
Tuesday: 6
Wednesday: 2
Thursday: 5
Friday: 4

What was the mean number of absences for the week?

A 3
B 4
C 5
D 6

43. **SDAP 1.3** The graph shows the average daily temperature this week. What is the difference between the hottest temperature and the coolest temperature?

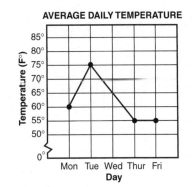

A 5°F
B 10°F
C 15°F
D 20°F

Benchmark Assessment

Name _____

PRACTICE TEST 2 • PAGE 8

Write the correct answer.

44. **SDAP 2.1** What is the number of different outcomes possible when tossing the quarter and spinning the pointer?

_____8_____

45. **MR 3.1** Raj spent 2 hours and 47 minutes at orchestra rehearsal. The rehearsal ended at 4:25 P.M. When did it begin?

_____1:38 P.M._____

Choose the letter of the correct answer.

46. **SDAP 2.2** What is the probability of rolling a 1, 2, 3, 4, or 6 on a number cube labeled 1–6?

   A $\frac{1}{6}$　　C $\frac{5}{6}$

   B $\frac{2}{6}$　　D Not Given

47. **MR 2.1** Brian bought some pens for $3.87, paper for $2.79, and a calculator for $9.95. Which is the best estimate of how much change he will receive from a $20 bill?

   A $3　　C $17
   B $16　　D $37

48. **MR 2.2** What is the area of the figure below?

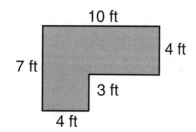

   A 52 sq ft　　C 106 sq ft
   B 82 sq ft　　D Not Given

49. **MR 1.2** Meg is planning a fun run. She needs to buy 154 T-shirts in each of 12 different colors. Which expression could she use to find the total number of T-shirts needed?

   A 154 + 12
   B (100 × 12) − (54 × 12)
   C (100 × 10) × (50 × 10) × (4 × 10) × 2
   D (100 × 12) + (50 × 12) + (4 × 12)

50. **MR 2.4** **Write About It** Explain how you solved Problem 45.

_____
_____
_____

BENCHMARK ASSESSMENT 1 SUMMARY AND PRESCRIPTIONS (1 of 4)

	California Standard	Test Items	Criterion for Success	Score	Harcourt Math PE/TE Lesson	Prescriptions for Intervention and Remediation
NS 1.0	Students understand the place value of whole numbers and decimals to 2 decimal places and how whole numbers and decimals relate to simple fractions. Students use the concepts of negative numbers.	5	1/1	__/1	1.3, 1.4	R-1.3, 1.4; P-1.3, 1.4; ATS: TE pp. 6B, 8B, 10; ISS: 1–4; EP: PE p. H32, Sets B–D
NS 1.1	Read and write whole numbers in the millions.	1, 2, 6	2/3	__/3	1.4	R-1.4; P-1.4; ATS: TE pp. 8B, 10; ISS: 1–4; EP: PE p. H32, Sets B–D
NS 1.2	Order and compare whole numbers and decimals to 2 decimal places.	7, 8	2/2	__/2	2.1, 2.2	R-2.1, 2.2; P-2.1, 2.2; ATS: TE pp. 18B, 20B, 22; ISS: 6; EP: PE p. H33, Sets A–B
NS 1.3	Round whole numbers through the millions to the nearest ten, hundred, thousand, ten thousand, or hundred thousand.	3, 4	2/2	__/2	2.4	R-2.4; P-2.4; ATS: TE pp. 26B, 28; ISS: 7; EP: PE p. H33, Set C
NS 1.4	Decide when a rounded solution is called for and explain why such a solution may be appropriate.	9, 11, 15	2/3	__/3	3.6	R-3.6; P-3.6; ATS: TE p. 48B; ISS: 10–12, 14, 18–20
NS 2.1	Estimate and compute the sum or difference of whole numbers and positive decimals to 2 places.	10, 12, 17	2/3	__/3	3.1	R-3.1; P-3.1; ATS: TE p. 34B; ISS: 8–12; EP: PE p. H34, Sets A–B
NS 3.0	Students solve problems involving addition, subtraction, multiplication, and division of whole numbers and understand the relationships among the operations.	16	1/1	__/1	8.1, 8.6	R-8.1, 8.6; P-8.1, 8.6; ATS: TE pp. 140B, 152B; ISS: 36–40; EP: PE p. H39, Sets A–C
NS 3.1	Demonstrate an understanding of, and the ability to use, standard algorithms for the addition and subtraction of multidigit numbers.	13, 14	2/2	__/2	3.3, 3.5	R-3.3, 3.5; P-3.3, 3.5; ATS: TE pp. 40B, 44B, 46; ISS: 10–12, 18–20; EP: PE p. H34, Sets C–E
NS 4.0	Students know how to factor small whole numbers.	20, 21	2/2	__/2	8.3	R-8.3; P-8.3; ATS: TE pp. 144B, 146; ISS: 36–40; EP: PE p. H39, Sets A–C

Key: R: Reteach P: Practice ATS: Alternative Teaching Strategy TE: Teacher's Edition ISS: Intervention Strategy Skill EP: Extra Practice PE: Pupil Edition

Benchmark Assessment

BENCHMARK ASSESSMENT 1 SUMMARY AND PRESCRIPTIONS (continued 2 of 4)

California Standard	Test Items	Criterion for Success	Score	Harcourt Math PE/TE Lesson	Prescriptions for Intervention and Remediation
AF 1.0 Students use and interpret variables, mathematical symbols, and properties to write and simplify expressions and sentences.	18, 19, 20, 21	3/4	__ /4	9.4, 9.5	R-9.4, 9.5; P-9.4, 9.5; ATS: TE pp.168B, 170B; ISS: 56–58; EP: PE p. H40, Sets C–F
AF 1.1 Use letters, boxes, or other symbols to stand for any number in simple expressions or equations (e.g., demonstrate an understanding and the use of the concept of a variable.)	22, 23	2/2	__ /2	9.5	R-9.5; P-9.5; ATS: TE p. 170B; ISS: 56–58; EP: PE p. H40, Sets C–F
AF 1.2 Interpret and evaluate mathematical expressions that now use parentheses.	26, 27	2/2	__ /2	9.1	R-9.1; P-9.1; ATS: TE pp. 158B, 160; ISS: 56–58; EP: PE p. H40, Set A
AF 1.3 Use parentheses to indicate which operation to perform first when writing expressions containing more than two terms and different operations.	34, 35	2/2	__ /2	4.2, 4.3, 9.1	R-4.2, 4.3, 9.1; P-4.2, 4.3, 9.1; ATS: TE pp. 56B, 58B, 158B, 160; ISS: 15–16, 56–58; EP: PE p. H35, Sets A–C; p. H40, Sets A–B
AF 1.5 Understand that an equation such as $y = 3x + 5$ is a prescription for determining a second number when a first number is given.	24, 28	2/2	__ /2	4.5, 9.6	R-4.5, 9.6; P-4.5, 9.6; ATS: TE pp. 64B, 172B; ISS: 53–58; EP: PE p. H35, Sets D–F; p. H40, Sets C–F
AF 2.0 Students know how to manipulate equations.	25, 29	2/2	__ /2	4.6, 9.3	R-4.6, 9.3; P-4.6, 9.3; ATS: TE pp. 66B, 164B, 166; ISS: 53-58; EP: PE p. H35, Sets D–F; p. H40, Sets C–F
AF 2.1 Know and understand that equals added to equals are equal.	30, 31	2/2	__ /2	4.6	R-4.6; P-4.6; ATS: TE pp. 66B, 68; ISS: 53–55; EP: PE p. H35, Sets D–F
AF 2.2 Know and understand that equals multiplied by equals are equal.	32, 33, 34, 35	3/4	__ /4	9.3	R-9.3; P-9.3; ATS: TE pp. 164B, 166; ISS: 56–58; EP: PE p. H40, Sets C–F

BENCHMARK ASSESSMENT 1 SUMMARY AND PRESCRIPTIONS (continued 3 of 4)

California Standard		Test Items	Criterion for Success	Score	Harcourt Math PE/TE Lesson	Prescriptions for Intervention and Remediation
SDAP 1.0	Students organize, represent, and interpret numerical and categorical data and clearly communicate their findings.	36, 37	2/2	__ /2	5.1, 5.6	R-5.1, 5.6; P-5.1, 5.6; ATS: TE pp. 82B, 84, 94B; ISS: 77–78; EP: PE p. H36, Sets A–C
SDAP 1.1	Formulate survey questions; systematically collect and represent data on a number line; and coordinate graphs, tables, and charts.	42, 43	2/2	__ /2	5.3, 6.2, 6.4	R-5.3, 6.2, 6.4; P-5.3, 6.2, 6.4; ATS: TE pp. 88B, 102B, 106B, 108; ISS: 77, 79; EP: PE p. H36, Sets A–C; p. H37, Set B
SDAP 1.2	Identify the mode(s) for sets of categorical data and the mode(s), median, and any apparent outliers for numerical data sets.	38, 39, 40	2/3	__ /3	5.2, 5.3	R-5.2, 5.3; P-5.2, 5.3; ATS: TE pp. 86B, 88B; ISS: 77; EP: PE p. H36, Sets A–C
SDAP 1.3	Interpret one- and two-variable data graphs to answer questions about a situation.	41	1/1	__ /1	5.3, 6.1, 6.2, 6.4	R-5.3, 6.1, 6.2, 6.4; P-5.3, 6.1, 6.2, 6.4; ATS: TE pp. 88B, 100B, 102B, 106B; ISS: 77, 79; EP: PE p. H36, Sets A–C; p. H37, Set B
MR 1.0	Students make decisions about how to approach problems.	47	1/1	__ /1	7.4, 8.6	R-7.4, 8.6; P-7.4, 8.6; ATS: TE pp. 124B, 152B; ISS: 23–24, 36–39, 62–63
MR 1.1	Analyze problems by identifying relationships, distinguishing relevant from irrelevant information, sequencing and prioritizing information, and observing patterns.	48	1/1	__ /1	7.3	R-7.3; P-7.3; ATS: TE pp. 120B, 122; ISS: 61–64; EP: PE p. H38, Sets C–D
MR 2.0	Students use strategies, skills, and concepts in finding solutions.	44	1/1	__ /1	1.5, 7.4	R-1.5, 7.4; P-1.5, 7.4; ATS: TE pp. 12B, 124B; ISS: 3–4, 62–63
MR 2.1	Use estimation to verify the reasonableness of calculated results.	45	1/1	__ /1	3.1	R-3.1; P-3.1; ATS: TE p. 34B; ISS: 8–12, 14, 17–20; EP: PE p. H34, Sets A–B
MR 2.3	Use a variety of methods, such as words, numbers, symbols, charts, graphs, tables, diagrams, and models, to explain mathematical reasoning.	49	1/1	__ /1	5.6, 6.5	R-5.6, 6.5; P-5.6, 6.5; ATS: TE pp. 94B, 110B; ISS: 78

Key: **R:** Reteach **P:** Practice **ATS:** Alternative Teaching Strategy **TE:** Teacher's Edition **ISS:** Intervention Strategy Skill **EP:** Extra Practice **PE:** Pupil Edition

BENCHMARK ASSESSMENT 1 SUMMARY AND PRESCRIPTIONS (continued 4 of 4)

California Standard	Test Items	Criterion for Success	Score	Harcourt Math PE/TE Lesson	Prescriptions for Intervention and Remediation
MR 2.6 Make precise calculations and check the validity of the results from the context of the problem.	50	1/1	__/1	2.3	R-2.3; P-2.3; ATS: TE p. 24B; ISS: 6
MR 3.1 Evaluate the reasonableness of the solution in the context of the original situation.	46	1/1	__/1	3.1	R-3.1; P-3.1; ATS: TE p. 34B; ISS: 8–12, 14, 17–20; EP: PE p. H34, Sets A–B

Key: **R:** Reteach **P:** Practice **ATS:** Alternative Teaching Strategy **TE:** Teacher's Edition **ISS:** Intervention Strategy Skill **EP:** Extra Practice **PE:** Pupil Edition

BENCHMARK ASSESSMENT 2 SUMMARY AND PRESCRIPTIONS (1 of 4)

	California Standard	Test Items	Criterion for Success	Score	Harcourt Math PE/TE lesson	Prescriptions for Intervention and Remediation
NS 1.2	Order and compare whole numbers and decimals to two decimal places.	1, 5	2/2	__ /2	19.4	R-19.4; P-19.4; ATS: TE pp. 366B, 368; ISS: 50–51; EP: PE p. H50, Sets B–C
NS 1.4	Decide when a rounded solution is called for and explain why such a solution may be appropriate.	2, 6	2/2	__ /2	10.2	R-10.2; P-10.2; ATS: TE p. 188B; ISS: 27–28; EP: PE p. H41, Sets B–D
NS 1.5	Explain different interpretations of fractions, for example, parts of a whole, parts of a set, and division of whole numbers; explain equivalents of fractions.	3, 9, 13	2/3	__ /3	17.1, 17.3	R-17.1, 17.3; P-17.1, 17.3; ATS: TE pp. 316B, 320B, 322; ISS: 45–46; EP: PE p. H48, Sets A–B
NS 1.6	Write tenths and hundredths in decimal and fraction notations and know the fraction and decimal equivalents for halves and fourths (e.g., $\frac{1}{2}$ = 0.5 or .50; $\frac{7}{4} = 1\frac{3}{4} = 1.75$).	4, 10	2/2	__ /2	19.1, 19.6	R-19.1, 19.6; P-19.1, 19.6; ATS: TE pp. 358B, 360, 372B, 374; ISS: 50–52; EP: PE p. H50, Sets A, D
NS 1.7	Write the fraction represented by a drawing of parts of a figure; represent a given fraction by using drawings; and relate to simple decimals on a number line.	7, 11	2/2	__ /2	17.1, 17.6, 19.1	R-17.1, 17.6, 19.1; P-17.1, 17.6, 19.1; ATS: TE pp. 316B, 330B, 358B, 360; ISS: 45–46, 50–52; EP: PE p. H48, Sets A–B; p. H50, Sets A, D
NS 1.9	Identify on a number line the relative position of fractions, mixed numbers, and positive decimals to two decimal places.	8, 12	2/2	__ /2	17.2, 19.4, 19.6	R-17.2, 19.4, 19.6; P-17.2, 19.4, 19.6; ATS: TE pp. 318B, 366B, 368, 372B; ISS: 45–46, 50–51; EP: PE p. H48, Sets A–B; p. H50, Sets B–C

Key: **R:** Reteach **P:** Practice **ATS:** Alternative Teaching Strategy **TE:** Teacher's Edition **ISS:** Intervention Strategy Skill **EP:** Extra Practice **PE:** Pupil Edition

Benchmark Assessment

BENCHMARK ASSESSMENT 2 SUMMARY AND PRESCRIPTIONS (continued 2 of 4)

	California Standard	Test Items	Criterion for Success	Score	Harcourt Math PE/TE lesson	Prescriptions for Intervention and Remediation
NS 2.0	Students extend their use and understanding of whole numbers to the addition and subtraction of simple decimals.	14, 16, 20	2/3	__/3	20.5	R-20.5; P-20.5; ATS: TE pp. 388B, 390; ISS: 21–23; EP: PE p. H51, Sets C–E
NS 2.1	Estimate and compute the sum or difference of whole numbers and positive decimals to two places.	15, 17	2/2	__/2	20.2	R-20.2; P-20.2; ATS: TE p. 382B; ISS: 21; EP: PE p. H51, Set B
NS 2.2	Round two-place decimals to one decimal or the nearest whole number and judge the reasonableness of the answer.	18	1/1	__/1	20.1, 20.2	R-20.1, 20.2; P-20.1, 20.2; ATS: TE pp. 380B, 382B; ISS: 21; EP: PE p. H51, Set A–B
NS 3.0	Students solve problems involving addition, subtraction, multiplication, and division of whole numbers and understand the relationships among the operations.	19, 21	2/2	__/2	15.5	R-15.5; P-15.5; ATS: TE p. 288B; ISS: 42–43
NS 3.2	Demonstrate an understanding of, and the ability to use, standard algorithms for multiplying a multidigit number by a two-digit number and for dividing a multidigit number by a one-digit number; use relationships between them to simplify computations and to check results.	22, 24	2/2	__/2	12.1, 14.2	R-12.1, 14.2; P-12.1, 14.2; ATS: TE pp. 220B, 222, 262B; ISS: 1–2, 30–33; EP: PE p. H43, Sets 30–32; p. H45, Sets A, B, D
NS 3.3	Solve problems involving multiplication of multidigit numbers by two-digit numbers.	23, 25	2/2	__/2	12.1, 12.2	R-12.1, 12.2; P-12.1, 12.2; ATS: TE pp. 220B, 222, 224B; ISS: 30–32; EP: PE p. H43, Sets 30–32
NS 3.4	Solve problems involving division of multidigit numbers by one-digit numbers.	26, 28	2/2	__/2	14.2, 14.4	R-14.1, 14.2, 14.4; P-14.1, 14.2, 14.4; ATS: TE pp. 260B, 262B, 268B; ISS: 1–2, 33; EP: PE p. H45, Sets A–B, D
NS 4.0	Students know how to factor small whole numbers.	27, 29	2/2	__/2	16.1, 16.2	R-16.1, 16.2; P-16.1, 16.2; ATS: TE pp. 294B, 296B; ISS: 24–25, 34; EP: PE p. H47, Sets A–D

© Harcourt

BENCHMARK ASSESSMENT 2 SUMMARY AND PRESCRIPTIONS (continued 3 of 4)

	California Standard	Test Items	Criterion for Success	Score	Harcourt Math PE/TE lesson	Prescriptions for Intervention and Remediation
NS 4.1	Understand that many whole numbers break down in different ways (e.g., $12 = 4 \times 3 = 2 \times 6 = 2 \times 2 \times 3$).	30, 32	2/2	__/2	16.2	R-16.2; P-16.2; ATS: TE p. 296B; ISS: 24-25, 34; EP: PE p. H47, Sets A–D
NS 4.2	Know that numbers such as 2, 3, 5, 7, and 11 do not have any factors except 1 and themselves and that such numbers are called prime numbers.	31, 33	2/2	__/2	16.3, 16.4	R-16.3, 16.4; P-16.3, 16.4; ATS: TE pp. 298B, 300, 302B; ISS: 24-25, 34; EP: PE p. H47, Sets A–D
AF 1.0	Students use and interpret variables, mathematical symbols, and properties to write and simplify expressions and sentences.	34, 36	2/2	__/2	10.6	R-10.6; P-10.6; ATS: TE p. 200B; ISS: 27
MR 1.0	Make decisions about how to approach problems.	35, 37	2/2	__/2	18.4, 19.5	R-18.4, 19.5; P-18.4, 19.5; ATS: TE pp. 348B, 370B; ISS: 47–48, 50; EP: PE p. H49, Set B
MR 1.1	Analyze problems by identifying relationships, distinguishing relevant information from irrelevant information, sequencing and prioritizing information, and observing patterns.	38, 42, 44	2/3	__/3	16.5	R-16.5; P-16.5; ATS: TE p. 304B; ISS: 24–25
MR 1.2	Determine when and how to break a problem into simpler parts.	39, 43	2/2	__/2	11.4	R-11.4; P-11.4; ATS: TE p. 212B; ISS: 27, 29; EP: PE p. H42, Sets A–B
MR 2.1	Use estimation to verify the reasonableness of calculated results.	40, 47	2/2	__/2	20.6	R-20.6; P-20.6; ATS: TE p. 392B; ISS: 21–23
MR 2.3	Use a variety of methods, such as words, numbers, symbols, charts, graphs, tables, diagrams, and models, to explain mathematical reasoning.	41, 48	2/2	__/2	17.5, 19.5	R-17.5, 19.5; P-17.5, 19.5; ATS: TE pp. 328B, 370B; ISS: 45, 50
MR 2.6	Make precise calculations and check the validity of the results from the context of the problem.	45, 49	2/2	__/2	16.5	R-16.5; P-16.5; ATS: TE p. 304B; ISS: 24–25

Key: **R:** Reteach **P:** Practice **ATS:** Alternative Teaching Strategy **TE:** Teacher's Edition **ISS:** Intervention Strategy Skill **EP:** Extra Practice **PE:** Pupil Edition

Benchmark Assessment

BENCHMARK ASSESSMENT 2 SUMMARY AND PRESCRIPTIONS (continued 4 of 4)

California Standard		Test Items	Criterion for Success	Score	Harcourt Math PE/TE lesson	Prescriptions for Intervention and Remediation
MR 3.1	Evaluate the reasonableness of the solution in the context of the original situation.	46, 50	2/2	__ /2	14.5, 20.6	R-14.5, 20.6; P-14.5, 20.6; ATS: TE pp. 270B, 392B; ISS: 1–2, 21–23

Key: **R:** Reteach **P:** Practice **ATS:** Alternative Teaching Strategy **TE:** Teacher's Edition **ISS:** Intervention Strategy Skill **EP:** Extra Practice **PE:** Pupil Edition

BENCHMARK ASSESSMENT 3 SUMMARY AND PRESCRIPTIONS (1 of 4)

	California Standard	Test Items	Criterion for Success	Score	Harcourt Math PE/TE lesson	Prescriptions for Intervention and Remediation
NS 1.8	Use concepts of negative numbers (e.g., on a number line, in counting, in temperature, in "owing").	1, 3, 5	2/3	__ /3	23.3	R-23.3; P-23.3; ATS: TE pp. 442B, 444; ISS: 5–6, 59; EP: PE p. H60, Set C
AF 1.0	Students use and interpret variables, mathematical symbols, and properties to write and simplify expressions and sentences.	2, 4, 6	2/3	__ /3	21.3, 22.2	R-21.3, 22.2; P-21.3, 22.2; ATS: TE pp. 410B, 426B; ISS: 29, 35; EP: PE p. H52, Set C; p. H53, Set B
AF 1.1	Use letters, boxes, or other symbols to stand for any number in simple expressions or equations (e.g., demonstrate an understanding and the use of the concept of the variable).	8	1/1	__ /1	21.3, 22.2	R-21.3, 22.2; P-21.3, 22.2; ATS: TE pp. 410B, 426B; ISS: 29, 35; EP: PE p. H52, Set C; p. H53, Set B
AF 1.4	Use and interpret formulas (e.g., area = length × width or $A = lw$) to answer questions about quantities and their relationships.	9, 11	2/2	__ /2	26.2, 26.5	R-26.2, 26.5; P- 26.2, 26.5; ATS: TE pp. 498B, 506B, 508; ISS: 70–73; EP: PE p. H57, Sets B–E
AF 1.5	Understand that an equation such as $y = 3x + 5$ is a prescription for determining a second number when a first number is given.	10	1/1	__ /1	24.3	R-24.3, 24.4; P-24.3, 24.4; ATS: TE pp. 458B, 460, 462B, 464; ISS: 60–61; EP: PE p. H55, Sets C–D
MG 1.0	Students understand perimeter and area.	14	1/1	__ /1	26.4	R-26.4; P-26.4; ATS: TE p. 504B; ISS: 70–73; EP: PE p. H57, Sets B–E
MG 1.1	Measure the area of rectangular shapes by using appropriate units, such as square centimeter (cm^2), square meter (m^2), square kilometer (km^2), square inches (in^2), square yard (yd^2), square mile (m^2).	16	1/1	__ /1	26.3	R-26.3; P-26.3; ATS: TE pp. 500B, 502; ISS: 70–73; EP: PE p. H57, Sets B–E
MG 1.2	Recognize that rectangles having the same area can have different perimeters.	17	1/1	__ /1	26.4	R-26.4; P-26.4; ATS: TE p. 504B; ISS: 70–73; EP: PE p. H57, Sets B–E

Key: **R:** Reteach **P:** Practice **ATS:** Alternative Teaching Strategy **TE:** Teacher's Edition **ISS:** Intervention Strategy Skill **EP:** Extra Practice **PE:** Pupil Edition

Benchmark Assessment

BENCHMARK ASSESSMENT 3 SUMMARY AND PRESCRIPTIONS (continued 2 of 4)

	California Standard	Test Items	Criterion for Success	Score	Harcourt Math PE/TE lesson	Prescriptions for Intervention and Remediation
MG 1.3	Understand that rectangles that have the same perimeter can have different areas.	19	1/1	__ /1	26.4	R-26.4; P-26.4; ATS: TE p. 504B; ISS: 70–73; EP: PE p. H57, Sets B–E
MG 1.4	Understand and use formulas to solve problems involving perimeters and areas of rectangles and squares. Use these formulas to find the areas of more complex figures by dividing them into parts with these basic shapes.	7	1/1	__ /1	26.3	R-26.3; P-26.3; ATS: TE pp. 500B, 502; ISS: 70–73; EP: PE p. H57, Sets B–E
MG 2.0	Students use two-dimensional coordinate grids to represent points and graph lines and simple figures.	12	1/1	__ /1	24.1, 24.4	R-24.1, 24.4; P-24.1, 24.4; ATS: TE pp. 452B, 462B, 464; ISS: 60–61; EP: PE p. H55, Sets A–D
MG 2.1	Draw the points corresponding to linear relationships on graph paper (e.g., draw 10 points on the graph of the equation $y = 3x$ and connect them by using a straight line).	18	1/1	__ /1	24.4	R-24.4; P-24.4; ATS: TE pp. 462B, 464; ISS: 60–61; EP: PE p. H55, Sets C–D
MG 2.2	Understand that the length of a horizontal line segment equals the difference of the x-coordinates.	15	1/1	__ /1	24.2	R-24.2; P-24.2; ATS: TE pp. 454B, 456; ISS: 60; EP: PE p. H55, Sets A–B
MG 2.3	Understand that the length of a vertical line segment equals the difference of the y-coordinates.	20	1/1	__ /1	24.2	R-24.2; P-24.2; ATS: TE pp. 454B, 456; ISS: 60; EP: PE p. H55, Sets A–B
MG 3.0	Students demonstrate an understanding of plane and solid geometric objects and use this knowledge to show relationships and solve problems.	21, 22	2/2	__ /2	25.5, 27.4	R-25.5, 27.4; P-25.5, 27.4; ATS: TE pp. 490B, 524B; ISS: 36, 68–69
MG 3.1	Identify lines that are parallel and perpendicular.	23	1/1	__ /1	25.2	R-25.2; P-25.2; ATS: TE p. 482B; ISS: 67; EP: PE p. H56, Sets A–B
MG 3.2	Identify the radius and diameter of a circle.	24	1/1	__ /1	28.3, 28.4	R-28.3, 28.4; P-28.3, 28.4; ATS: TE pp. 534B, 536B; ISS: 75; EP: PE p. H59, Set A

BENCHMARK ASSESSMENT 3 SUMMARY AND PRESCRIPTIONS (continued 3 of 4)

California Standard		Test Items	Criterion for Success	Score	Harcourt Math PE/TE lesson	Prescriptions for Intervention and Remediation
MG 3.3	Identify congruent figures.	25	1/1	__/1	25.3	R-25.3; P-25.3; ATS: TE pp. 484B, 486; ISS: 68–69; EP: PE p. H56, Sets C–D
MG 3.4	Identify figures that have bilateral and rotational symmetry.	30	1/1	__/1	25.4, 25.5	R-25.4, 25.5; P-25.4, 25.5; ATS: TE pp. 488B, 490B; ISS: 68–69; EP: PE p. H56, Sets C–D
MG 3.5	Know the definitions of a right angle, an acute angle, and an obtuse angle. Understand that 90°, 180°, 270°, and 360° degrees are associated, respectively, with $\frac{1}{4}$, $\frac{1}{2}$, and $\frac{3}{4}$ full turns.	27, 29	2/2	__/2	25.1, 28.1	R-25.1, 28.1; P-25.1, 28.1; ATS: TE pp. 478B, 480, 530B; ISS: 67, 75; EP: PE p. H56, Sets A–B
MG 3.6	Visualize, describe, and represent geometric solids (e.g., prisms, pyramids) in terms of the number and shape of faces, edges, and vertices; interpret two-dimensional representations of three-dimensional objects; and draw patterns (of faces) for a solid that, when cut and folded, will make a model of the solid.	26	1/1	__/1	27.1, 27.2	R-27.1, 27.2; P-27.1, 27.2; ATS: TE pp. 516B, 518, 520B; ISS: 71, 74, 75; EP: PE p. H58, Sets A–B
MG 3.7	Know the definitions of different triangles (e.g., equilateral, isosceles, scalene) and identify their attributes.	28	1/1	__/1	28.5	R-28.5; P-28.5; ATS: TE p. 538B; ISS: 75; EP: PE p. H59, Sets B–C
MG 3.8	Know the definition of different quadrilaterals (e.g., rhombus, square, rectangle, parallelogram, trapezoid).	31	1/1	__/1	28.6, 28.7	R-28.6, 28.7; P-28.6, 28.7; ATS: TE pp. 540B, 542, 544B; ISS: 75; EP: PE p. H59, Sets B–C
SDAP 1.1	Formulate survey questions; systematically collect and represent data on a number line; and coordinate graphs, tables, and charts.	32, 34	2/2	__/2	29.3	R-29.3; P-29.3; ATS: TE p. 560B; ISS: 82
SDAP 2.0	Students make predictions for simple probability situations.	33, 37	2/2	__/2	29.4	R-29.4; P-29.4; ATS: TE pp. 562B, 564; ISS: 82; EP: PE p. H60, Set B

Key: **R:** Reteach **P:** Practice **ATS:** Alternative Teaching Strategy **TE:** Teacher's Edition **ISS:** Intervention Strategy Skill **PE:** Pupil Edition **EP:** Extra Practice

Benchmark Assessment

BENCHMARK ASSESSMENT 3 SUMMARY AND PRESCRIPTIONS (continued 4 of 4)

	California Standard	Test Items	Criterion for Success	Score	Harcourt Math PE/TE lesson	Prescriptions for Intervention and Remediation
SDAP 2.1	Represent all possible outcomes for a simple probability situation in an organized way (e.g., tables, grids, tree diagrams).	36, 42	2/2	__ /2	29.1, 29.2	R-29.1, 29.2; P-29.1, 29.2; ATS: TE pp. 556B, 558B; ISS: 81–82; EP: PE p. H60, Set A
SDAP 2.2	Express outcomes of experimental probability situations verbally and numerically (e.g., 3 out of 4; $\frac{3}{4}$).	35, 43	2/2	__ /2	30.1	R-30.1; P-30.1; ATS: TE p. 570B; ISS: 45–46; EP: PE p. H61, Set A
MR 1.0	Students make decisions about how to approach problems:	13, 41	2/2	__ /2	27.4	R-27.4; P-27.4; ATS: TE p. 524B; ISS: 36
MR 1.1	Analyze problems by identifying relationships, distinguishing relevant from irrelevant information, sequencing and prioritizing information, and observing patterns.	38	1/1	__ /1	26.6, 30.4	R-26.6, 30.4; P-26.6, 30.4; ATS: TE pp. 510B, 576B; ISS: 45–46, 72–73
MR 2.3	Use a variety of methods, such as words, numbers, symbols, charts, graphs, tables, diagrams, and models, to explain mathematical reasoning.	39, 40, 44, 45	3/4	__ /4	21.6, 25.5	R-21.6, 25.5; P-21.6, 25.5; ATS: TE pp. 416B, 490B; ISS: 35, 65, 68–69
MR 2.5	Indicate the relative advantages of exact and approximate solutions to problems and give answers to a specified degree of accuracy.	48	1/1	__ /1	21.2	R-21.2; P-21.2; ATS: TE pp. 406B, 408; ISS: 35, 65; EP: PE pp. H52, Sets A–B
MR 2.6	Make precise calculations and check the validity of the results from the context of the problem.	49	1/1	__ /1	29.3	R-29.3; P-29.3; ATS: TE p. 560B; ISS: 82
MR 3.2	Note the method of deriving the solution and demonstrate a conceptual understanding of the derivation by solving similar problems.	46, 47	2/2	__ /2	21.6, 22.1	R-21.6, 22.1; P-21.6, 22.1; ATS: TE pp. 416B, 422B, 424; ISS: 29, 35, 65, 66
MR 3.3	Develop generalizations of the results obtained and apply them in other circumstances.	50	1/1	__ /1	23.4	R-23.4; P-23.4; ATS: TE p. 446B; ISS: 5–6

Key: **R:** Reteach **P:** Practice **ATS:** Alternative Teaching Strategy **TE:** Teacher's Edition **ISS:** Intervention Strategy Skill **EP:** Extra Practice **PE:** Pupil Edition

© Harcourt

CORRELATION OF CALIFORNIA MATHEMATICS STANDARDS TO *HARCOURT MATH* BENCHMARK ASSESSMENTS FOR GRADE 4

California Mathematics Standard	Correlation to *Harcourt Math* Benchmark Assessments
NUMBER SENSE	
NS 1.0 Students understand the place value of whole numbers and decimals to two decimal places and how whole numbers and decimals relate to simple fractions. Students use the concepts of negative numbers.	Benchmark Assessment 1: 5
NS 1.1 Read and write whole numbers in the millions.	Benchmark Assessment 1: 1, 2, 6 Practice Test 1: 1 Practice Test 2: 1
NS 1.2 Order and compare whole numbers and decimals to two decimal places.	Benchmark Assessment 1: 7, 8 Benchmark Assessment 2: 1, 5 Practice Test 1: 2 Practice Test 2: 2
NS 1.3 Round whole numbers through the millions to the nearest ten, hundred, thousand, ten thousand, or hundred thousand.	Benchmark Assessment 1: 3, 4 Practice Test 1: 3 Practice Test 2: 3
NS 1.4 Decide when a rounded solution is called for and explain why such a solution may be appropriate.	Benchmark Assessment 1: 9, 11, 15 Benchmark Assessment 2: 2, 6 Practice Test 1: 4 Practice Test 2: 4
NS 1.5 Explain different interpretations of fractions, for example, parts of a whole, parts of a set, and division of whole numbers; explain equivalents of fractions.	Benchmark Assessment 2: 3, 9, 13 Practice Test 1: 5 Practice Test 2: 5
NS 1.6 Write tenths and hundredths in decimal and fraction notations and know the fraction and decimal equivalents for halves and fourths (e.g., $\frac{1}{2} = 0.5$ or $.50$; $\frac{7}{4} = 1\frac{3}{4} = 1.75$).	Benchmark Assessment 2: 4, 10 Practice Test 1: 6 Practice Test 2: 6
NS 1.7 Write the fraction represented by a drawing of parts of a figure; represent a given fraction by using drawings; and relate to simple decimals on a number line.	Benchmark Assessment 2: 7, 11 Practice Test 1: 7 Practice Test 2: 7
NS 1.8 Use concepts of negative numbers (e.g., on a number line, in counting, in temperature, in "owing").	Benchmark Assessment 3: 1, 3, 5 Practice Test 1: 8 Practice Test 2: 8
NS 1.9 Identify on a number line the relative position of fractions, mixed numbers, and positive decimals to two decimal places.	Benchmark Assessment 2: 8, 12 Practice Test 1: 9 Practice Test 2: 9
NS 2.0 Students extend their use and understanding of whole numbers to the addition and subtractions of simple decimals.	Benchmark Assessment 2: 14, 16, 20
NS 2.1 Estimate and compute the sum or difference of whole numbers and positive decimals to two places.	Benchmark Assessment 1: 10, 12 Benchmark Assessment 2: 15, 17 Practice Test 1: 10 Practice Test 2: 10
NS 2.2 Round two-place decimals to one decimal or the nearest whole number and judge the reasonableness of the answer	Benchmark Assessment 2: 18, 22 Practice Test 1: 11, 13 Practice Test 2: 11, 13
NS 3.0 Students solve problems involving addition, subtraction, multiplication, and division of whole numbers and understand the relationships among the operations.	Benchmark Assessment 1: 16 Benchmark Assessment 2: 19, 21
NS 3.1 Demonstrate an understanding of, and the ability to use, standard algorithms for the addition and subtraction of multidigit numbers.	Benchmark Assessment 1: 13, 14 Practice Test 1: 12 Practice Test 2: 12

Benchmark Assessment

California Mathematics Standard	Correlation to *Harcourt Math* Benchmark Assessments
NUMBER SENSE (continued)	
NS 3.2 Demonstrate an understanding of, and the ability to use, standard algorithms for multiplying a multidigit number by a two-digit number and for dividing a multidigit number by a one-digit number; use relationships between them to simplify computations and to check results.	Benchmark Assessment 2: 22, 24 Practice Test 1: 14 Practice Test 2: 14
NS 3.3 Solve problems involving multiplication of multidigit numbers by two-digit numbers.	Benchmark Assessment 2: 23, 25 Practice Test 1: 15 Practice Test 2: 15
NS 3.4 Solve problems involving division of multidigit numbers by one-digit numbers.	Benchmark Assessment 2: 26, 28 Practice Test 1: 16 Practice Test 2: 16
NS 4.0 Students know how to factor small whole numbers.	Benchmark Assessment 1: 20, 21 Benchmark Assessment 2: 27, 29
NS 4.1 Understand that many whole numbers break down in different ways (e.g., $12 = 4 \times 3 = 2 \times 6 = 2 \times 2 \times 3$).	Benchmark Assessment 2: 30, 32 Practice Test 1: 17 Practice Test 2: 17
NS 4.2 Know that numbers such as 2, 3, 5, 7, and 11 do not have any factors except 1 and themselves and that such numbers are called prime numbers.	Benchmark Assessment 2: 31, 33 Practice Test 1: 18 Practice Test 2: 18
ALGEBRA AND FUNCTIONS	
AF 1.0 Students use and interpret variables, mathematical symbols, and properties to write and simplify expressions and sentences.	Benchmark Assessment 1: 18, 19, 20, 21 Benchmark Assessment 2: 34, 36 Benchmark Assessment 3: 2, 4, 6
AF 1.1 Use letters, boxes, or other symbols to stand for any number in simple expressions or equations (e.g., demonstrate an understanding and the use of the concept of a variable.)	Benchmark Assessment 1: 22, 23 Benchmark Assessment 3: 8 Practice Test 1: 19 Practice Test 2: 19
AF 1.2 Interpret and evaluate mathematical expressions that now use parentheses.	Benchmark Assessment 1: 26, 27 Practice Test 1: 20 Practice Test 2: 20
AF 1.3 Use parentheses to indicate which operation to perform first when writing expressions containing more than two terms and different operations.	Benchmark Assessment 1: 34, 35 Practice Test 1: 22 Practice Test 2: 22
AF 1.4 Use and interpret formulas (e.g., area = length $\times$ width or $A = lw$) to answer questions about quantities and their relationships.	Benchmark Assessment 3: 9, 11 Practice Test 1: 21 Practice Test 2: 21
AF 1.5 Understand that an equation such as $y = 3x + 5$ is a prescription for determining a second number when a first number is given.	Benchmark Assessment 1: 24, 28 Benchmark Assessment 3: 10 Practice Test 1: 23 Practice Test 2: 23
AF 2.0 Students know how to manipulate equations.	Benchmark Assessment 1: 25, 29
AF 2.1 Know and understand that equals added to equals are equal.	Benchmark Assessment 1: 30, 31 Practice Test 1: 24 Practice Test 2: 24
AF 2.2 Know and understand that equals multiplied by equals are equal.	Benchmark Assessment 1: 32, 33 Practice Test 1: 25 Practice Test 2: 25

California Mathematics Standard	Correlation to *Harcourt Math* Benchmark Assessments
MEASUREMENT AND GEOMETRY	
MG 1.0 Students understand perimeter and area.	Benchmark Assessment 2: 1, 2 Benchmark Assessment 3: 14
MG 1.1 Measure the area of rectangular shapes by using appropriate units, such as square centimeter (cm^2), square meter (m^2), square kilometer (km^2), square inches (in^2), square yard (yd^2), square mile (mi^2).	Benchmark Assessment 3: 16 Practice Test 1: 26 Practice Test 2: 26
MG 1.2 Recognize that rectangles having the same area can have different perimeters.	Benchmark Assessment 3: 17 Practice Test 1: 27 Practice Test 2: 27
MG 1.3 Understand that rectangles that have the same perimeter can have different areas.	Benchmark Assessment 3: 19 Practice Test 1: 28 Practice Test 2: 28
MG 1.4 Understand and use formulas to solve problems involving perimeters and areas of rectangles and squares. Use these formulas to find the areas of more complex figures by dividing them into parts with these basic shapes.	Benchmark Assessment 3: 7 Practice Test 1: 29 Practice Test 2: 29
MG 2.0 Students use two-dimensional coordinate grids to represent points and graph lines and simple figures.	Benchmark Assessment 3: 12
MG 2.1 Draw the points corresponding to linear relationships on graph paper (e.g., draw 10 points on the graph of the equation y = 3x and connect them by using a straight line).	Benchmark Assessment 3: 18 Practice Test 1: 30 Practice Test 2: 30
MG 2.2 Understand that the length of a horizontal line segment equals the difference of the x-coordinates.	Benchmark Assessment 3: 15 Practice Test 1: 31 Practice Test 2: 31
MG 2.3 Understand that the length of a vertical line segment equals the difference of the y-coordinates.	Benchmark Assessment 3: 20 Practice Test 1: 34 Practice Test 2: 34
MG 3.0 Students demonstrate an understanding of plane and solid geometric objects and use this knowledge to show relationships and solve problems.	Benchmark Assessment 3: 21, 22
MG 3.1 Identify lines that are parallel and perpendicular.	Benchmark Assessment 3: 23 Practice Test 1: 37 Practice Test 2: 37
MG 3.2 Identify the radius and diameter of a circle.	Benchmark Assessment 3: 24 Practice Test 1: 32 Practice Test 2: 32
MG 3.3 Identify congruent figures.	Benchmark Assessment 3: 25 Practice Test 1: 36 Practice Test 2: 36
MG 3.4 Identify figures that have bilateral and rotational symmetry.	Benchmark Assessment 3: 30 Practice Test 1: 35 Practice Test 2: 35
MG 3.5 Know the definitions of a right angle, an acute angle, and an obtuse angle. Understand that 90°, 180°, 270°, and 360° are associated, respectively, with $\frac{1}{4}$, $\frac{1}{2}$, and $\frac{3}{4}$, and full turns.	Benchmark Assessment 3: 27, 29 Practice Test 1: 33 Practice Test 2: 33

Benchmark Assessment

California Mathematics Standard	Correlation to *Harcourt Math* Benchmark Assessments
MEASUREMENT AND GEOMETRY (continued)	
MG 3.6 Visualize, describe, and represent geometric solids (e.g., prisms, pyramids) in terms of the number and shape of faces, edges, and vertices; interpret two-dimensional representations of three-dimensional objects; and draw patterns (of faces) for a solid that, when cut and folded, will make a model of the solid.	Benchmark Assessment 3: 26 Practice Test 1: 38 Practice Test 2: 38
MG 3.7 Know the definitions of different triangles (e.g., equilateral, isosceles, scalene) and identify their attributes.	Benchmark Assessment 3: 28 Practice Test 1: 40 Practice Test 2: 40
MG 3.8 Know the definition of different quadrilaterals (e.g., rhombus, square, rectangle, parallelogram, trapezoid).	Benchmark Assessment 3: 31 Practice Test 1: 41 Practice Test 2: 41
STATISTICS, DATA ANALYSIS, AND PROBABILITY	
SDAP 1.0 Students organize, represent, and interpret numerical and categorical data and clearly communicate their findings.	Benchmark Assessment 1: 36, 37
SDAP 1.1 Formulate survey questions; systematically collect and represent data on a number line; and coordinate graphs, tables, and charts.	Benchmark Assessment 1: 42, 43 Benchmark Assessment 3: 32, 34 Practice Test 1: 39 Practice Test 2: 39
SDAP 1.2 Identify the mode(s) for sets of categorical data and the mode(s), median, and any apparent outliers for numerical data sets.	Benchmark Assessment 1: 38, 39, 40 Practice Test 1: 42 Practice Test 2: 42
SDAP 1.3 Interpret one- and two-variable data graphs to answer questions about a situation.	Benchmark Assessment 1: 41 Practice Test 1: 43 Practice Test 2: 43
SDAP 2.0 Students make predictions for simple probability situations.	Benchmark Assessment 3: 34, 37
SDAP 2.1 Represent all possible outcomes for a simple probability situation in an organized way (e.g., tables, grids, tree diagrams).	Benchmark Assessment 3: 36, 42 Practice Test 1: 44 Practice Test 2: 44
SDAP 2.2 Express outcomes of experimental probability situations verbally and numerically (e.g., 3 out of 4; $\frac{3}{4}$).	Benchmark Assessment 3: 35, 43 Practice Test 1: 46 Practice Test 2: 46
MATHEMATICAL REASONING	
MR 1.0 Students make decisions about how to approach problems:	Benchmark Assessment 1: 47 Benchmark Assessment 2: 35, 37 Benchmark Assessment 3: 13, 41
MR 1.1 Analyze problems by identifying relationships, distinguishing relevant from irrelevant information, sequencing and prioritizing information, and observing patterns.	Benchmark Assessment 1: 48 Benchmark Assessment 2: 38, 42, 44 Benchmark Assessment 3: 38

California Mathematics Standard	Correlation to Harcourt Math Benchmark Assessments
MATHEMATICAL REASONING (continued)	
MR 1.2 Determine when and how to break a problem into simpler parts.	Benchmark Assessment 2: 39, 43 Practice Test 1: 49 Practice Test 2: 49
MR 2.0 Students use strategies, skills, and concepts in finding solutions.	Benchmark Assessment 1: 36, 37, 44
MR 2.1 Use estimation to verify the reasonableness of calculated results.	Benchmark Assessment 1: 45 Benchmark Assessment 2: 40, 47 Practice Test 1: 47 Practice Test 2: 47
MR 2.2 Apply strategies and results from simpler problems to more complex problems.	Practice Test 1: 48 Practice Test 2: 48
MR 2.3 Use a variety of methods, such as words, numbers, symbols, charts, graphs, tables, diagrams, and models, to explain mathematical reasoning.	Benchmark Assessment 1: 49 Benchmark Assessment 2: 41, 48 Benchmark Assessment 3: 39, 40, 44, 45
MR 2.4 Express the solution clearly and logically by using the appropriate mathematical notation and terms and clear language; support solutions with evidence in both verbal and symbolic work.	Practice Test 1: 50 Practice Test 2: 50
MR 2.5 Indicate the relative advantages of exact and approximate solutions to problems and give answers to a specified degree of accuracy.	Benchmark Assessment 3: 48
MR 2.6 Make precise calculations and check the validity of the results from the context of the problem.	Benchmark Assessment 1: 50 Benchmark Assessment 2: 45, 49 Benchmark Assessment 3: 49
MR 3.0 Students move beyond a particular problem by generalizing to other situations.	See MR 3.1–MR 3.3
MR 3.1 Evaluate the reasonableness of the solution in the context of the original situation.	Benchmark Assessment 1: 46 Benchmark Assessment 2: 46, 50 Practice Test 1: 45 Practice Test 2: 45
MR 3.2 Note the method of deriving the solution and demonstrate a conceptual understanding of the derivation by solving similar problems.	Benchmark Assessment 3: 46, 47
MR 3.3 Develop generalizations of the results obtained and apply them in other circumstances.	Benchmark Assessment 3: 50